A Manual of
BACTERIOLOGY

MAVEN BOOKS

A Manual of
BACTERIOLOGY

Herbert U. Williams

MAVEN BOOKS

Chennai   Trichy   Tirunelveli   New Delhi

MAVEN BOOKS

An Imprint of **MJP Publishers**

ISBN 978-93-88694-98-8 **Maven Books**

All rights reserved   No. 44, Nallathambi Street,
Printed and bound in India Triplicane, Chennai 600 005

MJP 842   © Publishers, 2020

Publisher : C. Janarthanan

Publisher's Note

The legacy of a country is in its varied cultural heritage, historical literature, developments in the field of economy and science. The top nations in the world are competing in the field of science, economy and literature. This vast legacy has to be conserved and documented so that it can be bestowed to the future generation. The knowledge of this legacy is slowly getting perished in the present generation due to lack of documentation.

Keeping this in mind, the concern with retrospective acquiring of rare books has been accented recently by the burgeoning reprint industry. Maven Books is gratified to retrieve the rare collections with a view to bring back those books that were landmarks in their time.

In this effort, a series of rare books would be republished under the banner, "Maven Books". The books in the reprint series have been carefully selected for their contemporary usefulness as well as their historical importance within the intellectual. We reconstruct the book with slight enhancements made for better presentation, without affecting the contents of the original edition.

Most of the works selected for republishing covers a huge range of subjects, from history to anthropology. We believe this reprint edition will be a service to the numerous researchers and practitioners active in this fascinating field. We allow readers to experience the wonder of peering into a scholarly work of the highest order and seminal significance.

Maven Books

PREFACE TO THE THIRD EDITION.

THE plan used in the preceding editions of this manual has been followed in the preparation of the present one. The only departures have been in the insertion of a short historical sketch and the freer use of references to original articles and reviews. It is hoped that these features will assist in arousing the interest of students. As far as possible, reference has been made to articles in American and English journals likely to be easy of access. Besides the ones just named, numerous additions have been made which the recent advances in our knowledge have rendered necessary. Most of the illustrations of apparatus are new. The photomicrographs also are new and original with a few exceptions noted in the text. It is probably needless to say that none of them were retouched. The writer is indebted to the Gratwick Laboratory of Buffalo for the use of its facilities in making these photographs.

BUFFALO, NEW YORK, August, 1903.

PREFACE TO THE SECOND EDITION.

ALTHOUGH there has been no lack of works on bacteriology, it seemed to the writer that there was still a field open for one which sought to give the portions essential to medical science in a concise manner. It is gratifying, therefore, that the first edition of this little book should have been exhausted so soon.

Whether wisely or not, it is a fact that many medical schools require their students to absorb an amount of knowledge that taxes the brain to the utmost. While such conditions remain, the need is urgent for presenting what is taught in the accessory branches in as condensed a form as is consistent with a clear understanding of their great fundamental principles. It is mastery of such principles, after all, which is the object of a course in bacteriology, for they are essential to a correct understanding of most of the other branches. After that has been accomplished, (including the applications of bacteriology to diagnosis), it must be admitted that other branches deserve a larger amount of the student's time. This may be said without meaning to minimize the importance of bacteriology in the training of a physician. In the opinion of the writer it is neither possible nor desirable that every graduate should be a trained bacteriologist. However, no instructor can hope to bring the principles above mentioned home to his classes except by laboratory work. Very little attempt has been made to outline the program of a laboratory course, as that will always need to be planned according to the circumstances under which it is given.

The purpose of this book is to give in the smallest possible space the facts which a physician must know, with some of those which it is desirable that he should know, and a little of that which he may learn if his needs or inclinations lead him to go further. It is acknowledged, however, that, in deference to precedent, this purpose has not been carried to its fullest extent. Much time has been spent on the index, in order to make the contents quickly accessible. It is a source of regret to the writer that the additions which the revision seemed to demand have made the present book a little larger than the first edition.

BUFFALO, NEW YORK, June, 1901.

PREFACE TO THE FIRST EDITION.

In this manual the writer has endeavored to describe the laboratory technique which the beginner must follow, and at the same time to give a concise summary of the facts in bacteriology most important to the physician. In preparing a work of this character, which claims to be nothing more than a compilation, the standard text-books were necessarily consulted freely. On account of the need for brevity it has, in most cases, been impossible to mention authorities.

The writer is glad to have this opportunity to acknowledge his obligation to the works of Sternberg, Flügge, Günther, Eisenberg, Abbott, W. H. Park, Muir and Ritchie, Vaughan and Novy, and McFarland; and to numerous papers by Professor Welch and others. It is thought that the chapters on Germicides, and Surgical Disinfection, by Drs. Thos. B. Carpenter and Marshall Clinton, will be useful not only for the information presented in them, but especially in correlating that information with the facts of bacteriology.

Buffalo, New York, October, 1898.

CONTENTS.

PART I.

BACTERIOLOGICAL TECHNIQUE.

CHAPTER I.

CHAPTER II.

CHAPTER III.

CHAPTER IV.

CHAPTER V.

CHAPTER VI.

CHAPTER VII.

CHAPTER VIII.

APPENDIX.

PATHOGENIC PROTOZOA.

INTRODUCTION.

ANYONE who has not himself worked in a bacteriological laboratory finds it difficult to form a vivid conception of what bacteria are like, because among the familiar animals and plants there are none with which a close comparison can be made. Of the common organisms, perhaps ordinary yeasts and moulds are most like the bacteria. Yeasts and moulds, as everyone knows, grow on bread, cheese, meat, syrups and the like. They flourish in moist and dark places, as do mushrooms, puffballs and the other fungi. All these fungi, appearing so different in some respects, are alike in one particular, which is the absence of the green color that we are apt to think of as being the essential feature of vegetation. Plants that are green owe their color to a substance called chlorophyll. Upon the properties of this substance one of the most fundamental facts in biology depends. Under the influence of sunlight, by means of chlorophyll, plants are able to use as food the carbon dioxide which is always present in the atmosphere in small amounts. Although carbon dioxide is one of the most simple and stable of compounds, the union of its component elements is broken by the plant, and they are employed in the formation of other much more complex and unstable compounds, such as starch and cellulose, which enter into the plant's structure. The work of plants, it will be noticed, is, in the main, precisely the reverse of that performed by animals. Animals take the unstable carbohydrates with high potential energy, such as starches and sugars, as food, and exhale the stable carbon dioxide from the lungs. At the same time the animal receives the benefit of the energy resulting from the oxidation of the carbohydrates, which may appear indirectly in the form of nervous or muscular activity or warmth.

Those plants that are devoid of chlorophyll are compelled to some extent to use the same kinds of food as animals. They are unable to decompose carbon dioxide (in most cases), and procure their nourishment from the dead or living bodies of other plants or animals. Since they have no chlorophyll, light is of no advantage to them, and is often a positive detriment. Bacteria contain no chlorophyll, and are usually classed with the fungi, which they resemble in their inability to decompose carbon dioxide and to use it as food.[1]

There is another well-known property, possessed by yeasts especially, which may be useful in explaining the work done by bacteria. It is a fact of every-day observation that, when yeasts grow in dilute solutions of sugar, alcohol and gas are formed. It not only appears that bacteria sometimes form alcohol and gas from sugar, but that with different kinds of bacteria and different kinds of food material a great number of substances are made, some of which are powerful poisons. In most of the diseases caused by bacteria such poisons are produced within the living body of the patient. The symptoms of the disease and the changes in the patient's body are due to these poisons, rather than to the direct action of bacteria.

The extreme smallness of the bacteria prevents us from seeing them as individuals without the aid of the microscope, although great numbers of them taken together may form a plainly visible mass or growth. When they are examined with the microscope they appear as little round,

[1] See Chapter I., Part II.

rod-shaped or curved bodies, which may be likened to so many periods, dashes and commas. It is at once perceived that each bacterium is an individual by itself, and that it consists of a single cell, not of an aggregation of cells, as do most of the common plants and animals.

Under favorable conditions bacteria may be seen to multiply, one organism being divided by a partition into two parts, which separate and become two new organisms. The process is called *fission.*

At times certain bacteria present little bright spots which enlarge, and from which the rest of the cell breaks away in fragments. The bright body that remains is called a *spore,* and has greater resisting power against injurious influences than has the fully developed organism. To this extent these spores are something like the seeds of higher plants. There are spores that can withstand boiling for hours, but fortunately that is not true, as far as we know, of the spores of any of the bacteria that produce disease. The earlier investigators observed the appearance of bacteria in nutrient infusions which they had endeavored to sterilize by heat. They looked upon this fact as indicating the possibility of *spontaneous generation,* and it furnished the chief support of that theory. Probably their fluids contained very resistant spores, and were in reality not sterile.

From these facts, a definition for bacteria may be formulated.

Bacteria (Greek βαχτήριον, meaning a little stick) *are extremely minute, unicellular plants, which have no chlorophyll, and which divide by fission.* They are sometimes called *schizomycetes.* In every-day language they are known as *microbes,* and also as *germs.* They are generally classed with the fungi. In some respects they seem quite closely related to the algæ or simplest green plants, and, on the other hand, they have strong points of likeness with some of the unicellular animals belonging to the infusoria.

Bacteria are divided into three great groups:

Micrococci, or cocci[1] (singular, coccus)—spherical forms.

Bacilli (sing., bacillus)—long and straight, or rod-shaped bacteria.

Spirilla (sing., spirillum)—consisting of spiral filaments like the turns of a corkscrew, or parts of spirals shaped like commas.

The extreme smallness of the bacteria is hard of comprehension. We may say, of most of them, that from 5,000 to 25,000 placed end to end would make a line about an inch in length. When one touches a growth of bacteria with the sterilized platinum wire and spreads the tiny portion that adheres to the wire upon a slip of glass, it is found upon examination with the microscope that the bacteria left on the glass may be compared to the stars in the sky, the grains of sand on the shore, or any of the other standards for numbers that are nearly beyond computation.

FIG. 1.

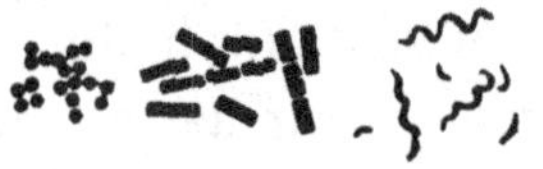

Micrococci. Bacilli. Spirilla.

It is well known that bacteria are present on most of the objects about us. They occur on the skins of men and other animals, as well as in the mouth, stomach and intestines, and on most of the surfaces of the body that open to the external world. They are found in the water of rivers and lakes, and in the ocean. They appear in the soil down to a depth of several feet. They float in the air, except at high altitudes and over the ocean. Nansen

[1] Pronounced kok´-sī or kok´-kē: see Webster's International, Century, and Standard Dictionaries, and Foster's and Keating's Medical Dictionaries. The writer knows of no authority for the prevailing pronunciation kok´-kī.

found bacteria on the ice of the Polar sea. Investigators have even reported finding them fossilized, indicating, as we might expect, that they existed at remote periods in the earth's history. But the vast majority of them are entirely harmless as far as we are concerned, and many of them are indispensable in maintaining the balance existing between the different kinds of living things.

Were it not for the putrefactive and nitrifying bacteria the dead bodies of plants and animals would lie practically unchanged where they fell, and the fertilization of the soil necessary for the life of most plants, by means of substances derived from such dead material, would cease.

In northern Siberia the bodies of the extinct species of elephant called mammoths have been found imbedded in frozen soil where they appear to have lain for thousands of years. In this case the growth of putrefactive bacteria has been prevented by cold, as in the modern refrigerator or cold-storage plant.

Some bacteria have been made to do work in industries, like the bacilli whose growth in cream imparts an agreeable flavor to the butter and cheese.

Bacteria are also made use of in the manufacture of vinegar.

The study of bacteria has led to the understanding of many hitherto unexplained facts. The unaccountable development of a moist, brilliant red deposit on bread and other articles of food, which was formerly believed by the superstitious to be blood, deposited by some miraculous agency, we know to be due to the growth of a common organism (bacillus prodigiosus). The emission of light by decaying substances when seen in the dark may be caused by bacteria as well as other organisms.

It seems that in some cases in which death was attributed to the suction of air into the veins, because air appeared to

Micrococci, for instance, which are, in reality, extremely different, may look very much alike. The differences are usually apparent when the bacteria are grown artificially. The cultivation is done for the most part in test-tubes containing some material which furnishes suitable food. The nutrient materials most used are meat-extract and peptone, which, dissolved with salt in water, constitute *nutrient bouillon.* Ordinary *gelatin,* or a vegetable gelatin called *agar-agar,* may be added to the bouillon when a solid *culture-medium* is desired. Before these substances can be used for the cultivation of bacteria all other bacteria which they might contain must be destroyed by heat.

When bacteria are to be conveyed from one tube to another, or from a tube to a glass slide, in order to examine them with the microscope, the manipulation is performed on a platinum wire fastened into a glass rod. The rules laid down for the management of the tubes and the platinum wire (Part I., Chapter VIII.) must be carefully followed. There is little or no danger in bacteriological work if the proper precautions are conscientiously observed; but carelessness may lead to disastrous and even fatal results, as has happened more than once.

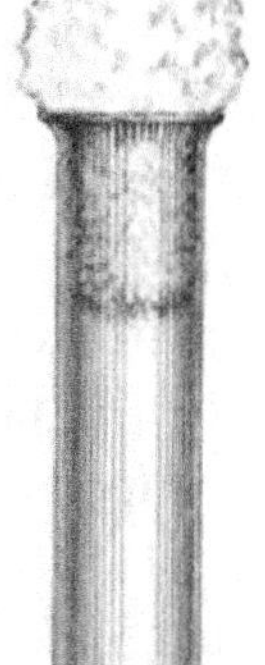

Fig. 2.

Test-tube containing culture-medium.

Finally, the effects of bacteria in bringing about disease may be tested on the lower animals. The proof that a particular species of bacteria causes a particular disease cannot be considered complete unless the disease can be reproduced by introducing these bacteria into some animal.

The student who wishes to pursue bacteriological study in any direction farther than it is possible for the limits of theories and fanciful speculations. But with all this we hear of certain beliefs and practices which plainly foreshadowed those of the present day. Latin writers nearly two thousand years ago recorded a relation between insects and malaria, which has but lately been proved and explained. The treatment of lepers by the Hebrews resembles that now in vogue: " He is unclean: he shall dwell alone; without the camp shall his habitation be " (Lev. XIII. 46). There is, in fact, much in the laws of Moses that points to some knowledge of the nature of infections. " This is the law, when a man dieth in a tent: all that come into the tent and all that is in the tent shall be unclean for seven days. And every open vessel which has no covering upon it shall be unclean" (Numb. XIX. 14, 15).

" Everything that may abide the fire, ye shall make it go through the fire, and it shall be clean " (Numb. XXXI. 23).

In Homer we read of Ulysses, that, having slain his wife's troublesome suitors:

" With fire and sulphur, cure of noxious fumes,
He purged the walls and blood-polluted rooms" (Pope's Odyssey).

The massive aqueducts of the Romans still remain to testify that they understood the importance of a pure water-supply.

In Rome there were also sewers for the disposal of drainage; while the Cretans and Assyrians used sewerage systems hundreds and even thousands of years before.

About the fourteenth century we find quarantine against infectious diseases, plague in particular, practiced by certain Italian cities; and the word " quarantine " came into use from the fact that the period of detention was about forty days (Ital. *quarantina*).[1]

[1] The Early History of Quarantine, J. M. Eager, *Yellow Fever Inst. Bul.*, No. 12, U. S. Marine Hosp. Service.

Leeuwenhoek, a citizen of Delft, in Holland (1632–1723), appears to have been the first who actually saw bacteria. Yeast-cells he certainly observed, besides making many other contributions of great value to biology. Leeuwenhoek produced admirable lenses of high magnifying power, and described what he witnessed with singular accuracy and enthusiasm.

Even before this time men had sought to explain the phenomena of infectious diseases by supposing the body to have been penetrated by minute parasites, for example worms. The spread of such diseases through a community from a single center could readily be accounted for by the multiplication of a contagious element, itself alive (*contagium vivum*). With increasing knowledge of the abundance of microscopic life these speculations took firmer hold. But long before their truth was finally demonstrated great advances were made in the prevention of infectious diseases. Much honor is due the clinicians whose accurate observations and foresight accomplished important results at an early day, working with what now seems a very meagre knowledge of the facts.

The production of immunity against small-pox by inoculation was first practiced in oriental countries. The method had long been in use in the East, when in 1718 it was brought to the notice of Europeans by Lady Montagu, wife of the English ambassador at Constantinople. The procedure consisted simply of the introduction of the virus of small-pox by puncture of the skin. An attack of small-pox resulted, which was much milder and far less dangerous than the natural disease.

Lady Montagu stated in a letter: " Every year thousands undergo the operation; and the French ambassador says pleasantly that they take the small-pox here by way of diversion, as they take the waters in other countries." The mild attacks that followed inoculation were, however, just as contagious to other persons as the natural disease, so that the dangers of this practice to the community were very great.

A much better method was found in vaccination. At this time a belief was current among farmers that a mild form of disease, called cow-pox, acquired by milkers, furnished protection against small-pox. This belief was investigated and introduced to the world by Edward Jenner. In 1796 he inoculated his first patient with cow-pox. In a few years the practice of vaccination spread to all parts of the world.[1]

It was introduced into the United States by Dr. Benjamin Waterhouse of Harvard. President Thomas Jefferson was active in bringing it into general use especially in the south.

As early as 1847 Semmelweis of Vienna attributed the origin of puerperal fever to poisons carried by the fingers of physicians and students, whose hands had been soiled in the dissecting room. To this he was led by the death of a friend from pyemia following a dissection-wound. He noted the similarity of the course of his friend's case with cases of puerperal fever. He advocated washing the hands of the attendant in solutions of chlorin or chlorid of lime, in addition to cleansing them with soap and water.

The cause of puerperal fever was still unknown. Endeavors to connect it with atmospheric influences and the like had been unsuccessful. During the seventeenth and eighteenth centuries it had been attributed to the absorption of milk into the blood from the breasts. Semmelweis stood his ground in spite of opposition and ridicule, though he somewhat modified his doctrine. His views agree substantially with the practice of the present day which they have greatly influenced.

[1] See the works of Edward Jenner by Dock, *N. Y. Med. Jour.*, Nov. 29 and Dec. 6, 1902; also The History of Vaccination, by Dulles, *Philadelphia Medical Journal*, May 30, 1903.

During the same period similar ideas were advanced by Dr. Oliver Wendell Holmes in the United States. His paper on " The Contagiousness of Puerperal Fever " appeared in 1843. A lively controversy lasting several years was provoked, in which Holmes defended his position with great vigor. His admirable literary style served him effectively.[1]

In the first half of the nineteenth century, with improved microscopes, knowledge of minute living things grew rapidly, chiefly with respect to infusoria and other relatively large forms. In 1840 Henle described the part played by microörganisms in producing disease in terms surprisingly in accord with views held at the present time. His deductions were based almost entirely on knowledge of the general nature, spread and course of infections. So too, Villemin anticipated the discovery of the bacillus of tuberculosis, for he transmitted the disease to animals, by inoculating them with material from cases of tuberculosis in man.

The key to exact knowledge of the microörganisms of disease was finally discovered in the study of fermentation. No better illustration could be found of the possible value to mankind which may lie in any addition whatever to the common stock of facts. The study of bottles of bad-smelling broth would have seemed, fifty years ago, a most unpromising beginning for the discovery of the causes of cholera, plague, and the like, or for an antitoxin for diphtheria.

Studies on Fermentation and Spontaneous Generation.—
Two observers (Schwann, Cagniard-Latour, 1837) almost simultaneously stated the proposition that yeast cells were living organisms, and that the fermentation of solutions of sugar was due to their growth. From this time ensued

[1] See Medical Essays, O. W. Holmes, Houghton, Mifflin & Co., 1889.

a controversy which lasted more than thirty years. The agency ascribed to yeasts was energetically denied by many, prominent among them Liebig; while it was sustained with vigor by others. The latter extended the original conception to include other sorts of fermentation and the putrefaction of albuminous material. Different kinds of fermentation, with different products, such as acetic acid and butyric acid, were ascribed to the growth of different kinds of microbes.

These microbes were found to be fungi of various sorts, and chiefly one or another variety of bacteria. The most celebrated among the students of fermentation was Pasteur, the simplicity and kindliness of whose character excite our admiration equally with his devotion to his work.[1]

Before the nature of fermentation was understood the possibility of spontaneous generation had been universally admitted. When vermin of various sorts appeared in putrefying material the conclusion was drawn that they had their origin directly from it. Although that had long since been disproved in the case of large organisms like worms and frogs, still, as late as the middle of the last century, it was held by many to account for the swarming microscopic life found in fermenting fluids. A flask of meat broth left exposed to the air will after a few days contain countless tiny living things, chiefly bacteria. Pasteur and his supporters showed that these bacteria were the progeny of others already in the flask or which had fallen in from the air.

When the flask of broth was boiled, no development of organisms took place, if the entrance of germs from the atmosphere was prevented. The latter was accomplished by such devices as heating the air, passing it through sulphuric acid, using a flask with a long twisted neck or by plugging the flask with cotton (Schröder and Von Dusch).

[1] See Louis Pasteur, His Life and Labors by His Son-in-Law, translated by Lady Claude Hamilton.

To prove that boiling had not made the fluid unfit for the growth of organisms, air was subsequently allowed to have access to it without such precautions, when putrefaction took place in the usual manner.

At the same time it was demonstrated not only that bacteria are present in all fermenting and putrefying substances, but that they exist wherever there is animal life or vegetation.

These principles underlie the methods used daily for the preservation of meat, fruit and vegetables, in the household and in factories.

Although even boiling occasionally failed to prevent fermentation, investigators came with practice to have a smaller number of failures. Such failures it was shown were due to the presence of the resistant state called spores, which some bacteria assume. The true nature of spores was recognized later by Cohn. Pasteur found that exposure to temperatures above the boiling point (110°C.) would destroy the most resistant microbes and their spores.

The controversies over fermentation and putrefaction lasted almost until the present day. They were productive of numerous benefits to the arts and manufactures. But what is of more importance to our subject, they led to a vastly better understanding of all kinds of microörganisms. The study of bacteria was now pursued with great vigor. In the space of about twenty-five years, most of what we know concerning the bacteria of disease has been learned. The period of rapid progress is not yet completed. Nearly every year yields some advance of great importance.

The discussions concerning fermentation and putrefaction, were still going on when Lister made his brilliant deduction that suppuration and septic processes in wounds were a species of fermentation (1867). From this came the antiseptic and aseptic methods of operating and of dressing wounds, which have made possible the wonderful results of modern operative surgery.[1]

In 1834 the parasite of itch (an insect, *Acarus scabei*) was discovered, and the cause of one contagious malady determined.

Quite early in the nineteenth century also the relatively large fungi of thrush and some of the parasitic skin diseases were discovered. The bacilli of anthrax, which are also large, were seen in the blood of animals by Pollender in 1855 and Davaine in 1863.

Davaine produced anthrax in animals by injecting into them blood containing anthrax bacilli. But complete proof that these bacilli were the cause of the disease, required that they should produce it when injected alone and when freed from the smallest trace of material derived from the first diseased animal. Unless these conditions were complied with, some other material, for example an enzyme or ferment, might be supposed to be carried from the first to the second animal and to be the real cause of the disease. For this purpose it was necessary to cultivate the bacilli in nutrient fluids, such as meat broth, as was done by Pasteur. It then became possible to demonstrate that their properties could remain unaltered after being grown in successive generations on different lots of broth. As bacteria of two or three species were often encountered in mixtures, it became most important to secure a method by which the different species could be separated from one another and be propagated as separate " pure cultures." This was done successfully by diluting such mixtures greatly, so that a drop planted in a new tube of broth should contain only a single organism. The growth ensuing would of course consist of the same kind of organism exclusively. Such procedures were uncertain and very laborious.

[1] See History of Medicine, Dr. Roswell Park.

Koch introduced in 1881 his method of separating bacteria by "plating" (described in Part I.), probably the most important single contribution to bacteriological technique. He also brought solid culture-media into general use by employing gelatin. Other important technical improvements of the same period were the adoption of the illuminating apparatus of Abbé and immersion objectives, and of aniline dyes for staining bacteria and making them visible (Weigert and Ehrlich). Beginning with the bacillus tuberculosis described by Koch in 1882, a large number of pathogenic bacteria were discovered during the ensuing years in rapid succession.

The application of the newly-gained knowledge concerning the bacteria causing infectious diseases to the prevention and cure of these diseases was begun almost immediately by Pasteur. A few facts existed to guide the direction of the research. It had been known even in ancient times that one attack of an infectious disease, such as scarlet fever, may confer immunity from subsequent attacks.

The protection against small-pox which was furnished by vaccination also was suggestive, although the mechanism by which this protection came about was not understood.

Pasteur worked on the theory that immunity to a disease might be secured by producing a mild attack of the disease. Such a mild attack might be expected to follow if a susceptible individual were inoculated with microbes of lowered virulence. Various methods were employed to reduce the virulence of bacteria, notably cultivation at high temperatures (43°C.). In this manner Pasteur was able to produce immunity against a number of the diseases of the lower animals. His method of inoculating sheep and cattle against anthrax is widely and successfully used. A similar principle has led to the preparation of a vaccine for the disease of cattle called "black leg," and such vaccine is now distributed gratuitously to farmers by the United States government. Inoculation of human subjects with the attenuated living virus of a disease is used only for hydrophobia. This method also was invented by Pasteur.

The preparations of antitoxins for infectious diseases (see the chapter on Immunity) we owe to Behring. This portion of our subject belongs entirely to the present day, and is now being studied with great energy.

Allusion has already been made to moulds and other microscopic parasites whose nature makes their study almost inseparable from that of the bacteria. In this class also belong the primitive forms of animal life (Protozoa) which are the causes of amebic dysentery (Lösch, 1875) and malaria (Laveran, 1880). The disease of cattle called " Texas fever " is also caused by a protozoön. Theobald Smith in the United States discovered that the parasite of Texas fever is conveyed from one animal to another by an insect, the cattle-tick. Since then it has been shown (by Manson, Ross and others) that malaria is conveyed from a person having the disease to one not affected by means of mosquitoes. It now appears probable that a similar relation exists between mosquitoes and yellow fever. The part played by flies and other insects in carrying disease germs is still receiving active attention and the future may have larger possibilities in store.

It is encouraging to reflect that the progress of bacteriology has been made by gradual and logical steps. The great discoveries have not been lucky accidents, but have been worked out patiently and with deliberation.

PART I.

CHAPTER I.

EXAMINATION OF BACTERIA WITH THE MICROSCOPE, IN-
CLUDING METHODS OF STAINING.

The Microscope.—The microscope consists of a tubular
body which carries the optical parts, and which can be
raised or lowered for focusing. The objectives should be
three in number, and should be attached to the body by
means of a triple nose-piece, which permits any objective
to be turned into the optical axis at will. The eye-piece
slips into the upper and opposite end of the body or tube.
The arrangements for focusing consist of a rack and pinion
which accomplish the coarse adjustment, and a more delicate fine adjustment. The stage, upon which the objects
to be examined are placed, has an opening in the middle.
In this opening an iris diaphragm and Abbé condenser are
inserted. The iris diaphragm enables one to alter the size
of the opening as desired. Beneath the stage is a movable mirror, of which one side is plane and the other concave. All of these parts are supported on a short, heavy
pillar which is fixed in the horseshoe-shaped base.

The essential parts of the microscope are, of course, the
eye-piece (German, *Ocular*), and the objective. Objectives
are given various names by different makers, for instance,
A, B, C, etc., or 1, 2, 3, etc.; or they are named according
to their focal distances, as $\frac{2}{3}$ inch, $\frac{1}{4}$ inch, $\frac{1}{2}$ inch, etc. In
bacteriological work a rather "low power" $\frac{2}{3}$ or $\frac{3}{4}$ inch

objective, an ordinary " high power " $\frac{1}{4}$ to $\frac{1}{6}$ inch dry objective, and a high power $\frac{1}{12}$ inch oil-immersion objective are needed. The magnification with the $\frac{2}{3}$ or $\frac{3}{4}$ inch objective

FIG. 3.

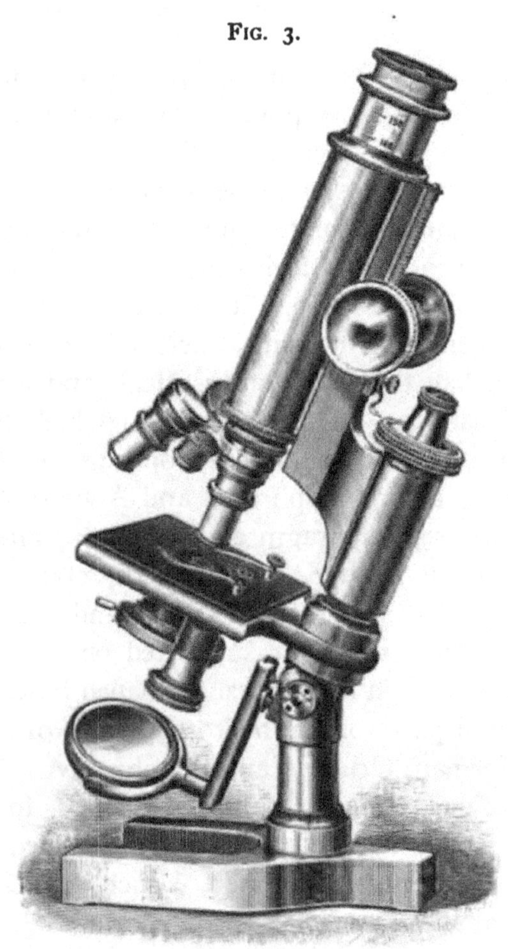

Microscope.

is about 75 to 100 diameters; with the $\frac{1}{4}$ to $\frac{1}{6}$ inch 300 to 500 diameters; with the $\frac{1}{12}$ immersion 750 to 1,000 diameters. The magnification varies according to the eye-piece used, as well as with the objective. A 1 inch and 1$\frac{1}{2}$ inch eye-piece (Zeiss No. 2 and No. 4) serve well for most purposes. The eye-pieces are usually named arbitrarily, like the objectives. The oil-immersion objective is used in the examination of bacteria where a very high power is desired. A layer of thickened oil of cedar-wood is placed between the lower surface of the objective and the upper surface of the glass covering the object under examination. The oil must be wiped away from the surface of the objective when the examination is finished. For this purpose the soft paper sold by dealers in microscopical apparatus serves admirably. Care must be taken not to scratch the lower surface of this objective. Oil of cedar-wood furnishes a medium

FIG. 4.

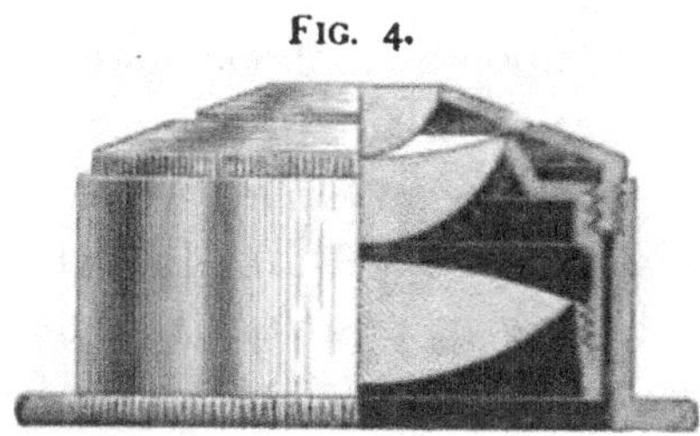

Abbé Condenser.   On the right side the figure gives a sectional view.

having nearly the same refractive index as the glass of the lens and the glass on which the object is mounted, and it obviates the dispersion of light which takes place when a layer of air is interposed between the objective and the object, as happens with the ordinary dry lens. This objective is used in connection with the Abbé condenser, which consists of two or three lenses combined so as to focus the rays coming from the plane mirror upon the object. The condenser gives a very intense illumination over a very small field. The condenser is not necessary excepting with the oil-immersion objective. If it is used with the other objectives the illumination must be regulated by lowering the condenser, closing the diaphragm more or less, and substituting the concave for the plane mirror. It is to be remembered that more depends upon securing a distinct picture than upon a very high magnification of the object.

The microscope should be placed in front of the observer on a firm table. The observer should be able to bring the eye easily over the eye-piece when the tube of the microscope is in vertical position. Daylight should be employed if possible. When artificial illumination is necessary, an ordinary lamp, a Welsbach burner or an incandescent electric light may be used. It is best to modify the artificial light by inserting a sheet of blue glass between the light and the mirror.

In order to focus upon any object, having first secured a satisfactory illumination with the mirror, it is best, beginning with the low power and using the coarse adjustment for focusing, to bring the objective quite close to the object, and then, with the eye in position, to raise the tube until the object comes into focus. The exact focusing is done with the fine adjustment. The observer should keep both eyes open when using the microscope, and should be able to use either eye at will.

All measurements of microscopic objects are expressed in terms of a micromillimeter. This is one-thousandth of a millimeter (.001 mm.), which is about $\frac{1}{25000}$ of an inch. It is generally called a micron for short, and is denoted by the Greek letter μ. For example, 5 μ = .005 mm. = $\frac{1}{5000}$ inch.

The Preparation of Specimens of Bacteria for Examination with the Microscope.—The substance under examination is usually placed upon thin slips of glass called cover-glasses. The material is spread over the cover-glass by means of a platinum wire which has been fixed in a glass rod about six inches long. Such a platinum wire is used constantly in doing bacteriological work. It is the tool by means of which one is able to handle bacteria with impunity. It serves in fact as a kind of additional finger. The platinum wire must be stiff enough not to bend too easily, and yet it should not be so large that it will not cool rapidly after heating. A good size for most purposes is number 23, American wire gauge (Brown and Sharp). The wire may be straight throughout its length, or the tip may be bent to form a loop (German, *Oese*). It is well to follow, from the beginning, certain rules which make the use of the platinum wire safe and accurate. Every time it is taken into the hand and before using it for any manipulation heat it in the flame of a Bunsen burner or

Fig. 5.

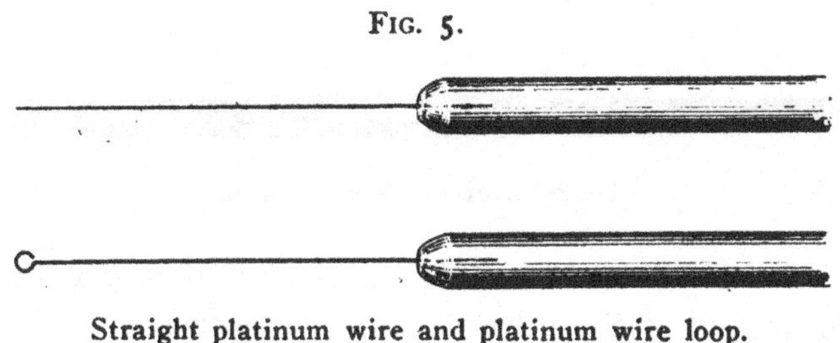

Straight platinum wire and platinum wire loop.

an alcohol lamp to a red heat; and always, *after using* and *before putting it down, heat it again to a red heat*. After the needle has become wet by dipping it in a fluid and is to be sterilized in the flame, it is necessary to avoid " sputtering " of the fluid by bringing the wet needle gradually to the flame, so as to dry the material adhering to it before burning it. This procedure must be done with great care when the wire has been dipped in milk or other substances containing oil. When the needle " sputters," as it is called, from too rapid heating, particles that have not yet been sterilized may be thrown some distance. On no account should the needle touch any object other than that which it is intended it should touch. With such a platinum wire, which has been properly sterilized, one can easily remove portions from a culture of bacteria, or from a fluid in which bacteria are supposed to be present. The glass rod in which the platinum wire is fixed should be held between the thumb and forefinger of the right hand like a pen. (For the manner of holding test-tubes, see page 84.)

The Hanging-drop.—Living bacteria may be studied with the microscope while suspended in some fluid substance. The needle having been heated to a red heat in the flame and having been allowed to cool, a small portion of the culture or other material may be removed with it and deposited in the center of an ordinary cover-glass. The needle should again be sterilized in the flame. When

FIG. 6.

Diagram of the hanging-drop.

cultures on solid media are to be examined, a small particle may be mixed with a drop of sterilized water or bouillon. The cover-glass should have been carefully cleaned and sterilized over the flame. The cover-glass with the small drop of fluid material held in sterilized forceps is now to be inverted over a sterilized glass slide, which has a concavity ground in the middle of it. Around the concavity, the slide should be smeared with vaseline. In this manner a small air-tight chamber is made. This slide and cover-glass may be put upon the stage of the microscope. A good dry lens, if of sufficiently high power, is more convenient for examining the hanging-drop than an oil-immersion. If the latter be used, having placed a drop of cedar-oil on the center of the cover-glass, and a good light having been secured, the oil-immersion objective should be brought down upon this drop of oil. The beginner often experiences difficulty in focusing upon a hanging-drop. It is well to shut off most of the light by means of the iris diaphragm. Often it is well to secure the focus roughly upon the extreme outer edge of the chamber, or to find the edge of the drop of fluid with the low power and then to focus upon this edge with the oil-immersion objective. Above all things guard against breaking the cover-glass by forcing the objective down upon it. The motility of certain bacteria is one of the most striking phenomena to be observed in the hanging-drop. It is not to be confused with the so-called " Brownian movement " which is exhibited by fine particles suspended in a watery fluid. It is well for the beginner to observe the character of the Brownian movement by rubbing up some carmine in a little water, and with the microscope to study the trembling motion exhibited by these particles of carmine. It will be noticed that, although the particles oscillate, no progress in any direction is accomplished unless there are currents in the fluid. Such currents might give rise to the impression that certain bacteria possessed motility when they were, in fact, powerless to move of themselves. In the hanging-drop the multiplication of bacteria can be studied, the formation of spores and the development of spores into fully formed bacteria. The hanging-drop has recently been put into service for the demonstration of the so-called serum-reaction with the bacillus of typhoid fever. Sometimes bacteria must be watched in the hanging-drop for hours, or even days, and it may be necessary to keep it at the temperature of the human body for this length of time. Various complicated kinds of apparatus have been devised for this purpose, but they are needful only with special kinds of work. When the hanging-drop preparation is no longer required, the slide and cover-glass should be dropped into a 5 per cent. carbolic acid solution and afterward sterilized by steam.

Hanging-block preparations, which were introduced by Hill,[1] make use of a cube of nutrient agar instead of a drop of fluid. Bacteria are distributed on the surface of the agar, which is then applied to a cover-glass, and mounted like a hanging-drop. The bacteria are kept in a layer close to the glass, where growth may be studied.

Cover-glass Preparations.—The study of bacteria with the microscope is for the most part done by means of smears made upon thin slips of glass. Such slips of glass are generally called cover-glasses. It is best to obtain the kind sold by dealers as No. 1, ¾ inch squares.

The cover-glass may be cleaned best by immersion in a mixture of sulphuric acid and bichromate of potassium solution, and afterward washed thoroughly in distilled water, and finally in alcohol. A stock of clean cover-glasses may be kept in a bottle of alcohol.

CLEANING FLUID.

Potassium bichromate...................... 40 grams.
Water .... 150 c.c.
Dissolve the bichromate of potassium in the
    water, with heat; allow it to cool; then add
    slowly and with care sulphuric acid, com-
    mercial .... 230 c.c.

For most purposes it is sufficient to wash the cover-glass in alcohol containing 3 per cent. of hydrochloric acid. It should then be wiped clean with a piece of linen cloth. Whenever it is taken into the fingers it should be held by the edges, never by the flat surfaces. As far as possible it should be handled with the forceps. It can be used very conveniently in the form of forceps known as the Cornet forceps, or in the modification devised by Stewart. Bac-

---

[1] *Journal of Medical Research,* Vol. VII., March, 1902.

teria may be placed upon the cover-glass by allowing the glass to fall upon one of the colonies of bacteria, on a gelatin or agar plate (see page 99), which will adhere to it in part, producing an " impression preparation " (German, *Klatsch-preparat*). Such a preparation, after drying in the air, is to be fixed by passing it through the flame three times. (See below.) The forceps with which it is handled should be sterilized in the flame.

Generally bacteria contained in fluids, like sputum, or taken from the surface of a culture, are smeared over the cover-glass by means of the platinum wire or loop, which must be heated to a red heat before and after the opera-

FIG. 7.

Cornet forceps for cover-glasses.

tion. Such preparations are called smear, cover-glass, cover-slip, or film preparations. When the material to be spread is thick or very viscid, a small drop of distilled water must first be placed in the center of the cover-glass so as to dilute it. Beginners generally take too much material on the wire. As thin a smear as possible is made. It is allowed to dry in the air; this should occupy a few seconds. The drying may be hastened by holding the forceps with the cover-glass a long distance above the flame, at a point where the heat would cause no discomfort to the hand. Having dried the preparation, it is to be passed through the flame of a Bunsen burner or alcohol lamp three times, taking about one second for each transit. The heat of the flame serves to dry the bacteria upon the cover-glass and fix them permanently in position; it is not

 MANUAL OF BACTERIOLOGY.

sufficient, however, when applied in this manner, to kill all kinds of bacteria, especially those containing spores. After it has been passed through the flame three times the preparation may be stained with one of the aniline dyes, and after washing in water and drying may be mounted,

FIG. 8.

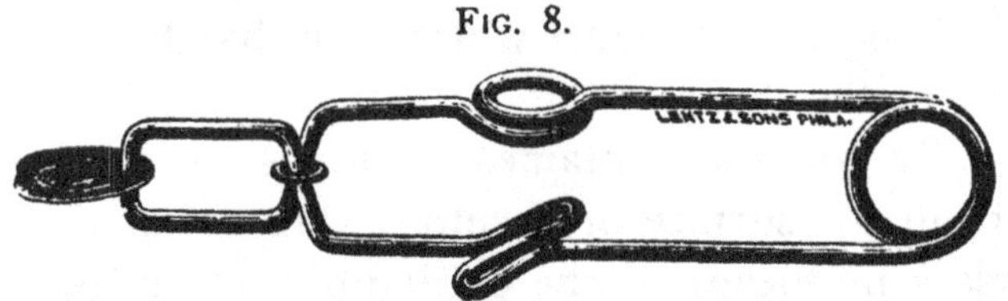

Stewart forceps for cover-glasses.

face down, in Canada balsam upon a glass slide. It makes a suitable object to be examined with the oil-immersion objective. The slide is a thin slip of glass, 3 inches by 1 inch, with ground edges.

The smear preparation may equally well be made directly upon the glass slide. The fixation in the flame must then occupy a longer time than with the small and thin cover-

FIG. 9.

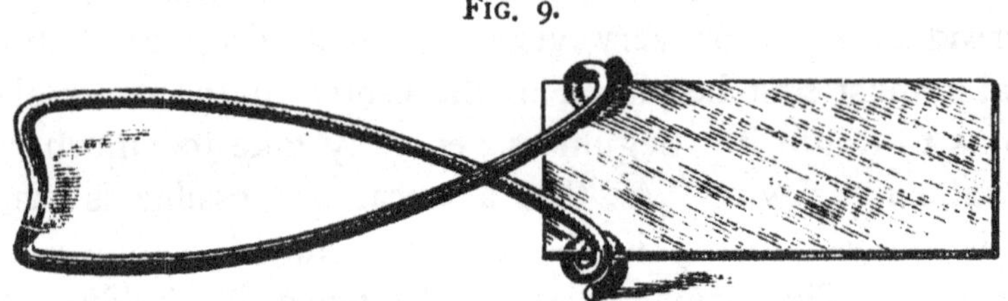

Kirkbride forceps for holding slides.

glass. Such preparations have the advantage that several may be made upon one slide, and that after staining them they may be examined in cedar-oil, with the oil-immersion lens, without the use of the cover-glass and Canada balsam. The forceps of Kirkbride will be found convenient when staining on the slide. Experiments performed in the writer's laboratory have shown that the ordinary method of fixation in the flame, when applied to bacteria spread upon slides, has little effect on the vitality of many species. The beginner is, therefore, advised to make his preparations on cover-glasses.

When very resistant or dangerous pathogenic bacteria are being handled, after fixation by heat upon the slide or cover-glass, the preparation may, if desired, be immersed in 1–1000 solution of bichloride of mercury long enough to kill the bacteria, without injuring the preparation or its staining properties.

Staining.—The staining of bacteria is done for the most part with the aniline dyes. The object of staining bacteria is to give them artificially some color which makes them distinct and easily visible without imparting this color to the substance or medium in which they are imbedded. The substances known as aniline dyes are derivatives of coal-tar, but not always of aniline. These dyes are of great importance in bacteriological work. Their number is very large, but only a few are in common use. It is important to have the purest, and those manufactured by Grübler are reliable.

It is simplest to classify the aniline dyes as acid or basic. Eosin, picric acid and acid fuchsin are acid dyes; they tend to stain tissues diffusely. Fuchsin, gentian-violet and methylene-blue are basic dyes; they have an affinity for the nuclei of tissues and for bacteria; they therefore are the dyes used chiefly in bacteriological work. The other varieties may be employed as contrast-stains; another contrast-stain frequently used is Bismarck brown. It is best to keep on hand saturated solutions of the aniline dyes in alcohol, from which watery solutions may be made when needed by adding a few drops of the alcoholic solution to a small dish filled with water. The alcoholic solution is diluted about ten times, or so as to make a liquid which is just transparent in a layer about 12 mm. in thickness, after filtering.

Fuchsin and gentian-violet operate rapidly and intensely. Methylene-blue works more slowly and feebly; it is to be preferred where the bacteria occur in thick or viscid substances, like pus, mucus, and milk.

Method of Staining Cover-glass Preparations.—(*a*) A smear preparation of bacteria having been made in the manner above described, and a watery solution of either fuchsin, gentian-violet or methylene-blue having been prepared, the cover-glass is to be dropped into a dish containing the dye, or the dye may be dropped upon the cover-glass held in the forceps.

(*b*) Allow the stain to act for about thirty seconds.

(*c*) Wash in water.

(*d*) Examine with the microscope in water directly or after drying and mounting in Canada balsam.

The rapidity and intensity of staining may be increased by warming the solution slightly. The bacteria will usually appear more distinct if, directly after pouring off the stain, the preparation is rinsed for a few seconds in 1 per cent. solution of acetic acid, and then thoroughly washed in water. The acetic acid solution serves to remove in a measure any color which has been imparted to the background, and which is undesirable.

Preparations that are mounted at first in water may be made permanent by moistening the edge of the cover-glass so that it may easily be removed from the slide, then drying and mounting in Canada balsam. Cover-glass preparations which have been stained are examined with the oil-immersion objective, employing the plane mirror, having the iris diaphragm open and the condenser close to the lower surface of the glass slide. The purpose is to obtain the most intense illumination possible over a small field.

The watery solutions of aniline dyes prepared as above described deteriorate in a short time, and it is best to prepare them freshly each time they are required. A very useful solution, which is permanent, is Löffler's alkaline methylene-blue:

<pre>
Concentrated alcoholic solution of methylene-blue..   30 c.c.
Potassium   hydrate   (caustic   potash)   1–10,000
    watery solution.................................... 100 c.c.
</pre>

Löffler's methylene-blue is a good stain for general purposes. It is perhaps more in use than any other formula for coloring the diphtheria bacillus.

Aniline-water Staining Solutions.—The intensity with which aniline dyes operate may be increased by adding aniline oil to the solution:

<pre>
Aniline oil ....   5 c.c.
Water .... 100 c.c.
</pre>

Mix, shake vigorously, filter; the fluid after filtration should be perfectly clear; add—

<pre>
Alcohol .... 10 c.c.
Alcoholic solution of fuchsin (or gentian-violet, or
    methylene-blue) .... 11 c.c.
</pre>

Aniline-water staining solutions do not keep well, and need to be freshly prepared about every two weeks. The applications of the aniline-water stains will be given under separate headings. In general, however, they are employed where a stain of unusual power is required.

Gram's Method.—Cover-glass preparations, having been prepared and fixed in the usual manner (see page 37), are stained as follows:

(*a*) Stain in aniline-water gentian-violet solution, from two to five minutes. The intensity of the stain may be increased by warming slightly.

(*b*) Iodine solution, one and one-half minutes:

Iodine .... 1 gram.
Potassium iodide .... 2 grams.
Water .... 300 c.c.

In this solution the preparation becomes nearly black.

(*c*) Wash in alcohol repeatedly; the alcohol becomes stained with clouds of violet coloring matter; the alcohol is used as long as the violet color continues to come away, and until the preparation is decolorized or has only a faint steel-blue color.

(*d*) When desired, the specimens may be stained, by way of contrast, with a watery solution of Bismarck brown or eosin.

(*e*) Wash in water, and examine either in water directly or after drying and mounting in Canada balsam. [A modification of this method, sometimes called the Gram-Günther method, differs from the preceding by using a 3 per cent. solution of hydrochloric acid in alcohol for ten seconds to hasten decolorization, washing in pure alcohol before and after the acid alcohol. Decolorization is more intense than by the Gram method; the diphtheria bacillus, which is stained by Gram's method, is decolorized by the Gram-Günther (Kruse).] The advantages of Gram's method are that with it certain bacteria are stained a violet color with more or less intensity and other bacteria are not stained at all. To some extent, then, it furnishes a means of diagnosis.

List of some of the important bacteria that are stained by Gram's method:

Staphylococcus pyogenes aureus,
Streptococcus pyogenes,
Micrococcus lanceolatus (of pneumonia),
Micrococcus tetragenus,
Bacillus of diphtheria,
Bacillus of tuberculosis,
Bacillus of leprosy,

Bacillus of anthrax,
Bacillus of tetanus,
Bacillus aërogenes capsulatus,
Ray fungus of actinomycosis.
The following bacteria are not stained by Gram's method:
Gonococcus,
Diplococcus intracellularis meningitidis,
Micrococcus melitensis,
Bacillus of chancroid (Ducrey),
Bacillus of dysentery (Shiga),
Bacillus of typhoid fever,
Bacillus coli communis,
Bacillus pyocyaneus,
Bacillus of influenza,
Bacillus of bubonic plague,
Bacillus of glanders (bacillus mallei),
Bacillus of malignant edema,
Bacillus of Friedländer,
Bacillus proteus,
Spirillum of Asiatic cholera,
Spirillum of relapsing fever.

Staining the Bacillus of Tuberculosis.—A very large number of methods have been proposed for staining the bacillus tuberculosis, all of which depend upon the principle that, after adding to solutions of aniline dyes certain substances, like aniline-water, carbolic acid, or solutions of ammonia or soda, the bacillus tuberculosis is stained with great intensity, and gives up its stain with difficulty. Solutions of acids will remove the stain from all parts of the preparation excepting from the tubercle bacilli, which retain the dye having once acquired it. The rest of the preparation may now be given a different color—contrast-stain.

Bacilli that resist decolorization by acids are called *acid-*

proof or acid-fast. The most important are tubercle and leprosy bacilli. There are various other species however most of which are less resistant to acids and alcohol than tubercle bacilli. They are discussed in the article on the bacillus tuberculosis in Part IV.

Occasionally spores of other bacteria, micrococci and horny epithelial cells are imperfectly decolorized, but their forms distinguish them from tubercle bacilli. Minute crystalline needles which have a shape like that of bacilli, are often encountered in sputum, but their nature will be recognized after a little practice.

The stain for tubercle bacilli is most frequently used for specimens of sputum from cases of suspected pulmonary tuberculosis; it may be applied to other fluids and secretions equally well. It is not reliable, however, when applied to milk, as the oil present in milk interferes with its operation, and milk and its products quite often contain other acid-proof bacilli. The smegma of the external genitals also frequently contains acid-proof bacilli that are not tubercle bacilli. On this account all fluids and discharges from the genito-urinary tract need to be examined with particular care not to confuse tubercle bacilli with smegma bacilli. (See smegma bacilli in Chapter IV., Part II.)

Patients should be given minute instructions concerning the collection of sputum. The bottle used should be new, wide-mouthed, clean, and kept tightly stoppered with a clean cork. The patient should be cautioned against allowing the expectoration to get on the outside of the bottle. Probably whatever risk is incurred by those who examine sputum comes chiefly from the outside of the bottle having been soiled with sputum containing tubercle bacilli. Often little white particles may be seen floating in the mucous portions of the sputum. These particles should be selected for the investigation, and may be spread in a thin film on the cover-glass with the platinum wire, which is sterilized in the flame before and after using. The selection of the little white particles will be facilitated if the sputum be poured into a clean glass dish, which may be placed on a black surface. A form of porcelain dish is furnished by dealers, the bottom of which is black, and which is convenient, for these manipulations. The smears must be made thin, or the subsequent decolorization, after staining, will not be uniform. It is hardly necessary to observe that the operator must be scrupulously careful not to contaminate the material under examination with any kind of extraneous matter. The cover-glasses and slides which are used should be new, and should have been cleaned with bichromate of potassium and sulphuric acid (see page 36).

When the work is completed, the bottle containing the sputum should be sterilized by steam or boiling.

Many different methods for staining the tubercle bacillus have been proposed. In most of those now in use the following solution is employed—

Fuchsin	1 gram.
Carbolic acid, pure	5 c.c.
Alcohol	10 c.c.
Distilled water	100 c.c.

The method given below is the one recommended.

Method for staining the tubercle bacillus:

(*a*) The cover-glass preparation is made, dried, and fixed by passing through the flame three times.

(*b*) The cover-glass, held in forceps or in a watch-crystal is covered with steaming carbol-fuchsin for five minutes.

(*c*) Wash in water.

(*d*) Wash in alcohol containing 3 per cent. of hydrochloric acid one minute, or longer if necessary to remove the red color.

(*e*) Wash in water.

(*f*) Stain with methylene-blue solution (see page 41) thirty seconds.

(*g*) Wash in water.

(*h*) Examine in water directly, or after drying and mounting in Canada balsam. Tubercle bacilli take a brilliant red color; other bacteria and the nuclei of cells are stained blue.

Gabbett's Method.—This method is very popular and widely used on account of its convenience. It is not as reliable as the one just given.

Gabbett's solution:

> Methylene-blue .... 1 to 2 grams.
> 25 per cent. watery solution of sulphuric acid. 100 c.c.

(*a*) The cover-glass preparation is to be made, dried, and fixed by passing through the flame three times.

(*b*) The carbol-fuchsin stain is applied from two to five minutes to the cover-glass, held in forceps or in a watch-crystal; it need not be warmed.

(*c*) Wash in water.

(*d*) Gabbett's solution is applied for one minute.

(*e*) Wash in water. The preparation should have a blue color. It may be examined in water directly or after drying and mounting in Canada balsam.

Gabbett's method has the advantage of decolorizing the preparation and staining the background with methylene-blue at the same time. Tubercle bacilli are colored a brilliant red; most other bacteria and the nuclei of cells are colored blue. The acid-proof bacilli mentioned on page 44 would keep the red stain also, in most cases, and would probably be confused with tubercle bacilli.

Of the numerous methods of staining tubercle bacilli only a few others can be mentioned. Aniline-water fuchsin, aniline-water gentian-violet, or carbol-fuchsin may be used.

The intensity of the stain must then be increased by warming the preparation till it steams or boils, then allowing the warm stain to act on the specimens for from three to five minutes; the preparation may also be left in the cold stain over night. Decolorization may be effected with a 25 per cent. solution of sulphuric acid used till the red color disappears, or a 30 per cent. solution of nitric acid, which operates very rapidly. If the red color persists after washing in water, dip in the acid again. After either acid the preparation is to be washed in alcohol until the last trace of the stain has been removed. An excellent decolorizing agent is a 3 per cent. solution of hydrochloric acid in alcohol, used for about a minute. With any of these acid solutions the decolorization can be accomplished more perfectly than with Gabbett's solution, where the operation of the decolorizing agent is masked. The contrast-stain may be omitted entirely if it is desired. A suitable contrast-stain after fuchsin staining is a solution of methylene-blue; after gentian-violet staining, Bismarck brown.

Those who have had experience in staining tubercle bacilli soon discover that the bacilli exhibit some differences in their resisting power to strong acids. One encounters occasionally bacilli that are perfectly stained side by side with others that are more or less completely decolorized. These facts show the necessity of practice with any method, and of exercising caution and judgment in making a diagnosis where the number of bacilli happens to be scanty. If tubercle bacilli are not found in the first preparation, other preparations should be made. Sometimes a large number of cover-glasses must be examined.

Various expedients have been devised to concentrate tubercle bacilli when only a small number may be present in a sample of sputum. In Biedert's method about 15 c.c. of sputum are mixed with 5 c.c. of distilled water, 4 to 8 drops of sodium hydrate solution are added, and the mixture is boiled. After boiling, add about 15 c.c. of distilled water. The mixture may be set aside in a conical glass for from twenty-four to forty-eight hours when the sediment may be collected, smeared on a cover-glass and stained for tubercle bacilli; or the sediment may be precipitated rapidly by the use of the centrifuge. The sediment will be found to have little adhesive power, and will not stick well to the cover-glass. It is convenient to save some of the original sputum and mix it with the sediment for this purpose.

Staining Bacteria in Tissues.—Pieces of organs about 1 cm. in thickness may be taken. Alcohol is the best agent for preserving them. The hardening will be completed in a few days. It is best to change the alcohol. The amount of the alcohol must be twenty times the bulk of the tissue to be preserved.

Ten parts of the standard 40 per cent. solution of formaldehyde, with 90 parts water make a good mixture for fixation; after twenty-four hours change to alcohol.

Imbedding in Collodion or Celloidin.—From alcohol the pieces of tissue are placed in equal parts of alcohol and ether twenty-four hours; thin collodion ($1\frac{1}{2}$ per cent.), twenty-four hours; thick collodion of a syrupy consistency (6 per cent.) twenty-four hours. The specimen is laid upon a block of wood and surrounded by thick collodion, and then inverted in 70 per cent. alcohol. The collodion makes a firm mass, surrounding and permeating the tissue, and permits very thin sections to be cut. The soluble cotton sold by dealers in photographer's supplies serves as well as the expensive preparation known as celloidin. To make collodion, dissolve it in equal parts of alcohol and ether. Soluble cotton is also called pyroxylin, and is a kind of gun-cotton.

Imbedding in Paraffin.—(*a*) Pieces of tissue 2 to 3 mm. thick which have already been fixed in alcohol or formaldehyde are to be placed in absolute alcohol for twenty-four hours.

(*b*) In pure xylol one to three hours.

(*c*) In a saturated solution of paraffin in xylol one to three hours.

(*d*) In melted paraffin having a melting-point of 50° C., which requires the use of a water-bath or oven, one to three hours. The xylol must be entirely driven off, and the tissue thoroughly infiltrated.

(*e*) Change to fresh paraffin for one hour.

(*f*) Finally, place the tissue in a small dish or paper box and pour the melted paraffin about it. Harden as quickly as possible with running water. It is important to fix the piece of tissue in a suitable position, if the position is of importance, before pouring in the melted paraffin.

Sections of exquisite thinness may now be cut. The knife need not be wet. Paraffin imbedding is especially desirable when serial sections are to be made.

In order to mount the sections, proceed as follows:

(*a*) Place the sections on water in a porcelain capsule. Warm slightly, when the sections will flatten nicely. Smear the surface of a slide with a very thin layer of Mayer's glycerin-albumen mixture. Dip the slide under the sections; lift them; and then drain off the water, leaving the sections in their proper positions. Let them dry for some hours in the incubator, and they will be firmly fastened to the slide.

(*b*) Dissolve out the paraffin in one of the numerous solvents (xylol, a few minutes).

(*c*) At this point the xylol should be washed off with absolute alcohol, and

(*d*) The section is stained.

(*e*) Dehydrate in absolute alcohol.
(*f*) Clear in xylol.
(*g*) Mount in balsam.

GLYCERIN-ALBUMEN MIXTURE (MAYER).

Equal parts of white of egg and glycerin are thoroughly mixed, and then filtered. Add a little gum-camphor to preserve.

Section Cutting.—Cutting is best done with an instrument called a microtome. The tissues may be imbedded in collodion or paraffin; or when they have been hardened with formaldehyde they may be cut after freezing. Bacteria

FIG. 10.

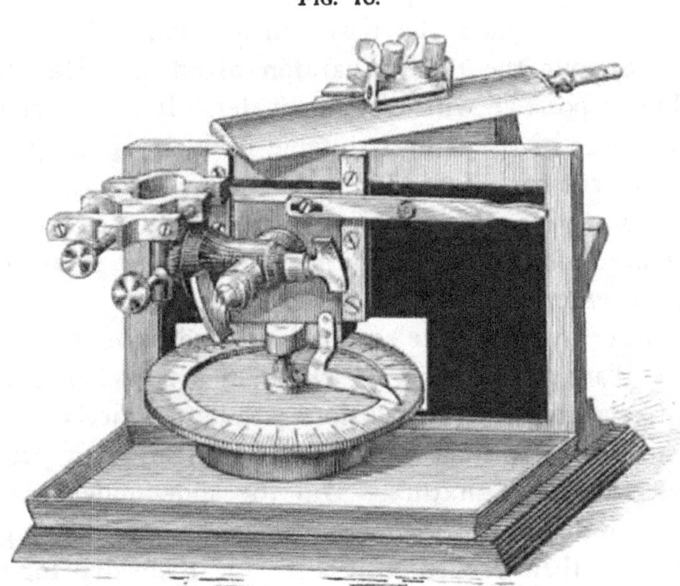

Schanze microtome.

stain admirably in frozen sections. For routine work collodion imbedding will be found as convenient a process as any. Paraffin imbedding gives the thinnest sections.

A microtome consists of a heavy, sliding knife-carrier, which moves with great precision on a level, and of a device for elevating the object which is to be cut any desired distance after each excursion of the knife. The thickness of the section will be the distance which the object is elevated. The knife is kept wet with alcohol during the cutting of collodion sections, otherwise it is left dry. The microtome is usually provided with a special form of knife. A razor will serve nearly as well, after having had the lower side ground flat. If a razor is used, a special form of razor-holder must be attached to the microtome to receive the razor. Above all, it is necessary that the knives should be kept in good condition. Only occasionally will they need honing, using a fine water-stone or Belgian hone. The movement in honing should be from heel to toe, always placing the back of the knife next the hone when turning. The knife should be stropped frequently. The leather of the strop should be glued to a strip of wood to make a flat surface. The movement in stropping should be from toe to heel. Sections should be cut to a thickness of not more than $25\,\mu$. Thinner sections (5 to $10\,\mu$) are to be desired.

Staining of Sections.—A watery solution of one of the aniline dyes is used—fuchsin, gentian-violet or methylene-blue—made by adding a few drops of the alcoholic solution to a dish filled with water. Löffler's solution of methylene-blue serves very well.

By this process most bacteria are stained; also the nuclei of cells; frequently, also, certain granules contained within some cells (German, *Mastzellen*), which may easily be mistaken for bacteria by the inexperienced (basophilic granules).

(*a*) Place the section in the staining solution from two to five minutes.

(*b*) Wash in water.

(*c*) Place in a watery solution of acetic acid, .1 per cent., for one minute.

(*d*) Alcohol, one to two minutes; change to absolute alcohol. Touch the sections to blotting-paper to remove the superfluous alcohol.

(*e*) Xylol until clear; xylol is to be preferred to other clearing agents, like oil of cloves, most of which slowly remove aniline colors. It has the disadvantage of not clearing when the slightest trace of water is present; dehydration in alcohol must, therefore, be complete. The section should be removed from the xylol as soon as it is cleared; otherwise wrinkling occurs.

(*f*) The section is placed upon a glass slide; a drop of Canada balsam is placed upon it and then a cover-glass. The Canada balsam should be dissolved in xylol.

The section is to be manipulated with straight or bent needles. The removal from xylol to the glass slide is managed best with a spatula or section-lifter.

The above statements apply to frozen sections or to sections imbedded in celloidin. Paraffin sections are preferably attached to the slide with glycerin-albumen. The different steps in the process follow in the same order. The stain may be poured on the slide, or the slide may be placed in a large dish full of staining fluid. (See page 49.) Celloidin sections may also be stained on the slide. If the section be well spread and flattened thoroughly with blotting-paper, it will usually adhere to the slide, and is less likely to wrinkle. It must not be allowed to dry.

Gram's Method may be applied to the staining of sections of tissues as well as to smears upon cover-glasses.

(*a*) Place the section in aniline-water gentian-violet, one to five minutes.

(*b*) Rinse briefly in water.

(*c*) Iodine solution (see page 42), one and one-half minutes.

(*d*) Alcohol, until decolorized to a faint blue-gray.

(*e*) Xylol.

(*f*) Mount on a slide in balsam.

Weigert's Modification of Gram's Method, or Weigert's Stain for Fibrin.—(*a*) Place the section in aniline-water gentian-violet solution, five minutes or more.

(*b*) Wash briefly in water.

(*c*) Place the section upon a slide by means of a section-lifter; having straightened it carefully, absorb the water with blotting-paper.

(*d*) Iodine solution (see page 42) one to two minutes.

(*e*) Absorb the iodine solution with blotting-paper.

(*f*) Add aniline oil, removing it from time to time with blotting-paper, and adding fresh aniline oil until the color ceases to come away. (Aniline oil serves in this connection both to decolorize and to dehydrate. It absorbs the water rapidly and efficiently. However, on account of its decolorizing tendency, it must be removed before the specimens can be mounted permanently.)

(*g*) Add xylol; remove it with blotting-paper; and add fresh xylol several times, in order to extract the last trace of aniline oil.

(*h*) Mount in Canada balsam.

This method is more convenient for the staining of sections than the Gram method. The results, however, are essentially the same as far as the bacteria are concerned; fibrin and hyaline material are stained blue, bacteria violet. It is often impossible to decolorize the nuclei completely without decolorizing the bacteria also. The parts of the nuclei which remain stained often present pictures that resemble bacteria, and which may lead to error if not recognized. Basophilic granules also retain the stain, as do the horny cells of the epidermis. These remarks apply also to Gram's method, except as regards fibrin. Very beautiful preparations can be obtained according to this or the Gram method when the sections have previously been stained in carmine; the nuclei will then be colored red, bacteria violet.

Tubercle bacilli may be *stained in sections* as follows:

(*a*) Use carbol-fuchsin, or aniline-water gentian-violet for one-half to two hours with very gentle warming, or over night without warming.

(*b*) Wash in water.

(*c*) Decolorize with some one of the decolorizing agents mentioned in connection with the staining of tubercle bacilli in cover-glass preparations, preferably 3 per cent. hydrochloric acid alcohol. Decolorization must be continued until the red color has disappeared, which requires one-half to several minutes.

(*d*) Wash in alcohol.

(*e*) Wash in water.

(*f*) Use hematoxylin as a contrast-stain for fuchsin preparations, and carmine for gentian-violet preparations. (It is better to stain with carmine first of all and before staining the bacilli. The carmine is not affected by the subsequent treatment.)

(*g*) Wash in water.

(*h*) Alcohol.

(*i*) Xylol.

(*j*) Balsam.

Nuclear stains, which may be used as contrast-stains for sections:

DELAFIELD'S HEMATOXYLIN.

Hematoxylin crystals	4 grams.
Alcohol	25 c.c.
Ammonia alum	50 grams.
Water	400 c.c.
Glycerin	100 c.c.
Methyl alcohol	100 c.c.

Dissolve the hematoxylin in the alcohol, and the ammonia alum in the water. Mix the two solutions. Let the mixture stand four or five days uncovered; it should have become a deep purple. Filter and add the glycerin and the methyl alcohol. After it has become dark enough, filter again. Keep it a month or longer before using; the solution improves with age. At the time of using, filter and dilute with water as desired.

LITHIUM-CARMINE (ORTH).

Carmine .... 2.5 grams.
Saturated watery solution of lithium carbonate. 100.0 c.c.

Add a few crystals of thymol. The carmine dissolves readily in the lithium carbonate solution. Filter the stain at the time of using. Sections are to be left in the stain five to twenty minutes.

Sections stained in carmine are placed directly in acid alcohol (1 part hydrochloric acid, 100 parts 70 per cent. alcohol) for five to ten minutes. They acquire a brilliant scarlet color. When used as a contrast-stain for tissues containing bacteria, it is best to use it before staining the bacteria, which might be decolorized by the acid alcohol.

Staining of Blood-Films.—The method of *Wright* is the one recommended. It is applicable to bacteria and to the parasite of malaria, and is useful as a general stain for blood. Films of blood are prepared as directed in chapter VII., Part I., and are allowed to dry.

(*a*) The stain is poured over the surface of the preparation till it covers it. This serves to fix the film of blood. It is allowed to remain for one minute.

(*b*) Add distilled water, drop by drop, till a reddish tint appears at the edges and a metallic scum forms on the surface. About six drops are needed for a three-fourths inch cover-glass. The real staining of the preparation now takes place, and requires two or three minutes.

(*c*) Wash in distilled water till the thin parts of the preparation have a yellowish or pinkish tint, which requires one to three minutes.

(*d*) Dry with blotting-paper and mount in Canada balsam.

Bacteria, malarial parasites, and cell-nuclei are stained blue, red blood-corpuscles are orange-pink, while the specific granules of the leucocytes (neutrophilic, etc.) appear in various tints from red to dark blue. The chromatin of the malarial parasite takes a lilac to red color. The blood-plates have a bluish or purplish color and must not be confused with malarial parasites.

The staining fluid is prepared as follows: To 100 c.c. of a one per cent. solution of sodium bicarbonate in water add 1 gram of methylene-blue. Place in the steam sterilizer at 100°C. for one hour. When cool add one-tenth per cent. watery solution of eosin (Grübler, yellowish, soluble in water) until the mixture loses its blue color, becomes purple, and a metallic scum forms on the surface. About 500 c.c. of the eosin solution are needed. Collect the precipitate on a filter; let it dry; make a saturated solution of the precipitate in methyl alcohol; filter. To the quantity obtained add one-fourth as much methyl alcohol, so that the solution may not be completely saturated. The purpose of the above procedures is to modify the methylene-blue so that other staining elements are developed in it (polychromism). The modified methylene-blue solution is then combined with eosin. For full details see Wright, *Journal of Medical Research,* Vol. VII. 1902.

Staining of Spores.—The method is applicable to cover-glass preparations which may be prepared in the usual way from material supposed to contain spores.

(*a*) After drying the smear on the cover-glass, and fixation with heat by passing through the flame three times, use as a stain aniline-water fuchsin.

(*b*) Heat until the preparation begins to boil; remove for a minute; heat again, and again remove; repeat this process six times.

(*c*) Wash in 3 per cent. hydrochloric acid alcohol one minute, or less.

(*d*) Wash in water.

(*e*) Stain with watery solution of methylene-blue half a minute.

(*f*) Wash.

(*g*) Dry.

(*h*) Balsam.

The spores are intensely stained by the fuchsin. The stain is removed from everything except the spores by the acid alcohol. The methylene-blue solution stains the bodies of the bacteria, the spores remaining brilliant red. There are various other methods for staining spores, but this procedure gives good results. The principle is the same as in staining the tubercle bacillus, except that more pains are needed to impregnate spores with the dye.

Staining of Capsules.—The capsules which many bacteria possess, appear to be made of some gelatinous substance, which is difficult to stain.

Method of Welch.—(*a*) Cover-glass preparations are made in the usual manner. Pour glacial acetic acid over the film.

(*b*) After a few seconds, replace with anilin-water gentian-violet, without washing in water. Change the stain several times to remove all the acetic acid. Allow it to act three or four minutes.

(*c*) Wash and examine in salt solution, 0.8 to 2.0 per cent.

Bacteria are deeply stained, while their capsules are pale violet. This method has been recommended for staining the capsule of the pneumococcus.

Methods of Hiss.—1. (*a*) Cover-glass preparations are made in the usual manner, and fixed in the flame.

(*b*) Stain for a few seconds in a half-saturated watery solution of gentian-violet.

(*c*) Wash in 25 per cent. solution of potassium carbonate in water.

(*d*) Mount and study in the same.

2. (*a*) Cover-glass preparations are made and fixed in the ordinary way.

(*b*) Use the following stain, heated till it steams:

Saturated alcoholic solution of gentian-violet or fuchsin. 5 c.c.
Distilled water .... 95 c.c.

(*c*) Wash in 20 per cent. solution of cupric sulphate crystals.

(*d*) Dry and mount in Canada balsam.

The methods of Hiss are recommended to be used for bacteria that have been cultivated on serum-agar with 1 per cent. of dextrose. They have shown that many streptococci have capsules. The writer has had good success from the latter method, with preparations of the pneumococcus from animal tissues.

Staining of Flagella.—Flagella are among the most difficult of all objects to stain. The best-known method is that of *Löffler*. It is important to use young cultures, preferably on agar.

(*a*) A small portion of the culture is mixed on a cover-glass with a drop of water. The preparations must be *exceedingly thin*. The mixing must be done with care in order not to break off the delicate flagella. The cover-glass must be perfectly clean, see page 36.

(*b*) After drying, fixation is effected by passing through the flame three times.

(*c*) The essential point in this method is the use of a mordant as follows:

Tannic acid, 20 per cent. solution................ 10 c.c.
Saturated solution of ferrous sulphate .... 5 c.c.
Saturated alcoholic solution of fuchsin.......... 1 c.c.

This solution is filtered and a few drops are placed on the cover-glass, or the cover-glass is placed, face down, in a dish containing the stain; it is then left for one to five minutes, warming slightly.

(*d*) Wash in water.

(*e*) Stain with aniline-water fuchsin, or carbol-fuchsin.

(*f*) Wash in water.

(*g*) Dry.

(*h*) Mount in Canada balsam.

(According to Löffler, certain bacteria require the addition of an acid solution, and certain others an alkaline solution, but many observers consider this unnecessary.)

Another and very valuable method is that of *Van Ermengem.*

(*a*) Make and fix cover-glass preparations as in the preceding method.

(*b*) Use the following mordant for one-half hour at room-temperature or for five minutes at 50° to 60° C.

Osmic acid 2 per cent. solution............................ 1
Tannic acid 10 to 25 per cent. solution.................. 2

(*c*) Wash carefully in distilled water and then in alcohol.

(*d*) Place for a few seconds in a 0.25 to 0.50 per cent. solution of nitrate of silver—" the sensitizing bath."

(*e*) Without washing transfer to the " reducing and reinforcing bath ":

Gallic acid .... 5 grams.
Tannic acid .... 3 grams.
Fused potassium acetate .... 10 grams.
Distilled water .... 350 c.c.

(*f*) After a few seconds, replace the preparation in the nitrate of silver solution, in which it is kept constantly moving, till the solution begins to acquire a brown or black color.

Some recommend leaving the preparation in the nitrate of silver solution for two minutes in the first place, and in

6

the reducing bath for two minutes, without using the nitrate of silver solution a second time.

(*g*) Finally wash in distilled water, dry, mount in Canada balsam. It is difficult to avoid the formation of precipitates; otherwise the results of this method are usually good.

CHAPTER II.

STERILIZATION.

By sterilization is meant the killing of all microörganisms found on or in any body or substance. It is possible to sterilize objects by the use of bichloride of mercury (corrosive sublimate), carbolic acid and other chemical agents. Sterilization is usually accomplished by heat. The most effective sterilization is that done by steam and by boiling; they are not, however, suitable for all kinds of material.

The naked flame of the Bunsen burner or the alcohol lamp is used largely for the sterilization of small articles. It is evident that no more efficient way of sterilization could be devised than by burning objects, or subjecting them to a red heat. The uses of this method will at once suggest themselves; for instance, surgical dressings that have become soiled with discharges and similar materials can be most easily disposed of by simply burning them up. In laboratory work the flame is constantly employed for the sterilization of the platinum wire, forceps, pipettes and cover-glasses; occasionally test-tubes are sterilized in this manner.

Hot-Air Sterilization.—Hot air, at a temperature of 150° C., or higher, maintained for an hour, is very valuable for some materials although less effective than steam. It has been found that the spores of certain bacteria are not killed even by exposure to this temperature, but it is sufficient for ordinary conditions. Hot-air sterilization is employed for glassware such as Petri dishes, flasks and test-tubes. Flasks and test-tubes are generally plugged with raw cotton. The sterilization should change the cotton to a light brown color, but it should not be scorched to a dark brown. Glassware should be placed within the sterilizer when it is cold, and after heating should be allowed to cool gradually in order to avoid breaking. Hot-air sterilization is never used for culture-media.

The hot-air sterilizer is a box made of sheet-iron, the walls being double, with an air-space between them. On

FIG. 11.

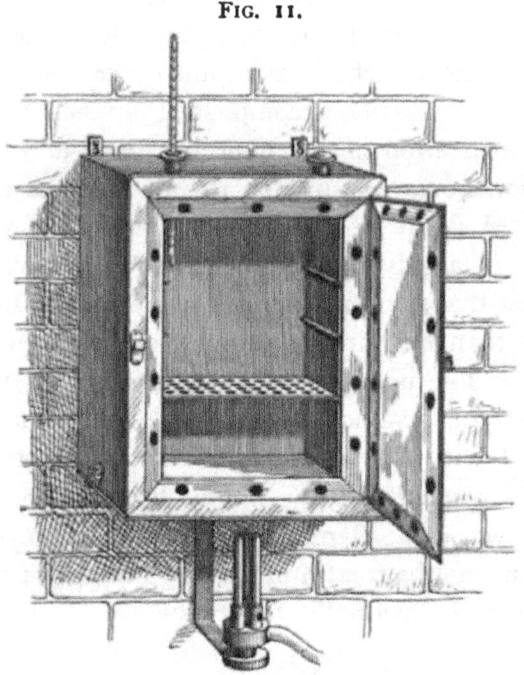

Hot-air sterilizer.

one side is a door. There are openings at the top to secure the circulation of air in the air-chamber. A thermometer passes from the top into the interior of the sterilizer so that one may read off the temperature that is being attained. The sterilizer should be placed so that there will be no danger of its setting fire to inflammable articles, as the heat may occasionally become very intense. It is well, if possible, to have it fastened to a brick wall.

Boiling.—Boiling is an efficient method of sterilization. It is often used for instruments. In laboratory work steam is generally substituted for it.

Steam Sterilization.—Steam sterilization is the most generally used of all forms of sterilization and is the most effective. It is employed for perishable bodies which would be injured by dry air sterilization or by chemical germicides; for example, it is used for surgical instruments and for culture-media; in laboratory work, especially for culture-media. It has been found that there are some forms of bacteria which, in the resting or spore stage, can resist even the action of steam for several hours. Such prolonged exposure to steam would be very injurious to culture-media, which are more or less unstable organic substances. What is called *fractional, intermittent* or *discontinuous* sterilization is used for such materials. By that plan the medium is sterilized with steam for fifteen minutes on each of three consecutive days. The object of intermittent sterilization as explained by Tyndall, who proposed it, is this: The culture-medium may be supposed to contain fully developed bacteria, and also bacteria in the spore or resting stage. The first sterilization of fifteen minutes will probably be sufficient to destroy all the fully developed bacteria; during the twenty-four hours between the first and second sterilization all of the spores which have survived the first sterilization may be expected to have become fully developed into bacteria which can be destroyed by the second sterilization; the third sterilization is directed against any spore forms which may possibly have survived the second sterilization.

Although the spore forms which are so extremely resistant are mostly non-pathogenic, as for example the bacilli of hay and potato, they nevertheless are capable of ruining the culture-media with which one works.

It has been shown by T. Smith that the discontinuous method cannot be relied upon to sterilize fluids in shallow layers that are freely exposed to the air. For if the spores of anaërobic bacteria happen to be present in such fluids, they will not develop into the adult form between the applications of heat, under aërobic conditions.

The form of sterilizer most widely used in the United States is that which is known as the Arnold Steam Sterilizer.

FIG. 12.

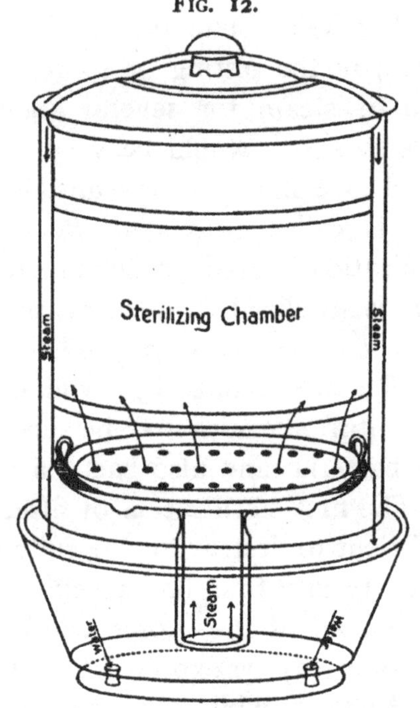

Diagram of the Arnold steam sterilizer.

The Arnold sterilizer consists of a cylinder of tin or copper with a cover, which is enclosed in a movable, cylindrical outer cover or hood. The inner cylinder has an opening in the bottom through which steam may enter, the steam coming from a small chamber underneath with a copper bottom to which the flame is applied. The peculiarity of this form of sterilizer consists in the fact that the steam which escapes from the sterilizing chamber will be condensed beneath the outer cover or hood and will fall back upon the pan over the chamber in which the steam is generated. The bottom of this pan is perforated with three

FIG. 13.

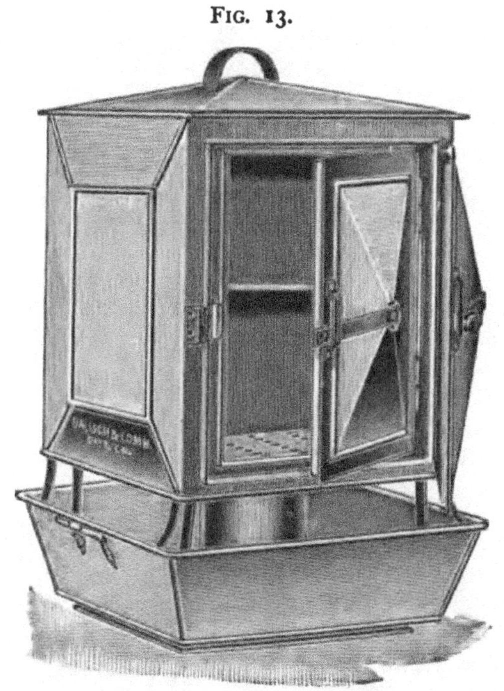

Steam sterilizer, Massachusetts Board of Health.

small holes which allow the water of condensation to return into the chamber where the steam is generated. The sterilizer will, therefore, to a certain extent, supply itself with water, although not by any means perfectly. It is, however, less likely to boil dry than other forms of sterilizers, and it has the advantage of being reasonably cheap and quite effective. The space enclosed by the hood also serves as a steam-jacket and helps to overcome fluctuations in temperature. A great improvement upon the ordinary Arnold sterilizer is the modification of it devised by the Massachusetts Board of Health.

In the use of this, or any form of steam sterilizer, the time when sterilization is supposed to begin must be counted

FIG. 14.

Koch's Steam Sterilizer.

as that when boiling is brisk and it is evident that the sterilizing chamber is filled with hot steam; or, what is better, when the thermometer registers 100° C., if the sterilizer be provided with a thermometer. With a large Arnold sterilizer a temperature of 100° C. may not be reached until it has been heated with a rose-burner for twenty to thirty-five minutes.

The sterilizer invented by Koch is still largely in use. It is a tall cylindrical tin vessel covered with asbestos or felt. The lower portion is filled with water; on the side is a water-gauge indicating the height of the water, in order that one may observe when there is danger of the sterilizer boiling dry. Over the top there is a tight-fitting cover. The steam is generated by a Bunsen burner standing underneath. A perforated shelf placed some distance above the surface of the water is for the reception of the tubes and flasks that are to be sterilized.

The *sterilization* of *blood-serum* sometimes has to be performed in a specially devised sterilizer, when a clear, fluid medium is desired. In this case the serum is heated for an hour on each of six consecutive days to a temperature of only 58° C. To obtain a transparent but solid medium the serum is kept at a temperature of 75° C. for an hour on each of four consecutive days. The process must be conducted carefully to avoid clouding of the serum.

Pasteurization.—The name pasteurization has been applied to the partial sterilization of substances at a comparatively low temperature. It is employed particularly for milk. The temperature used (70° to 75° C. for 20 to 30 minutes) is sufficient to destroy all ordinary pathogenic bacteria; for example, the bacilli of tuberculosis and typhoid fever. Furthermore, the great majority of the saprophytic bacteria are destroyed, and milk which has been pasteurized will remain unchanged for several days, if kept cool. Its application is principally in the feeding of infants when ordinary milk has been found to produce undesirable results. Freeman[1] has invented a pail of special form for the pasteurization of milk in bottles. This pail is filled with hot

[1] *Medical Record,* July 2, 1892, and August 4, 1894.

water and the bottles are placed in it; it has been found to keep up a temperature of about 75° C.

The Autoclave.—The autoclave is an instrument designed for sterilization by steam under pressure. It was invented in France but is now used extensively in all parts of the world. Steam generated at the ordinary atmospheric pressure is much less destructive to bacteria, and especially

FIG. 15.

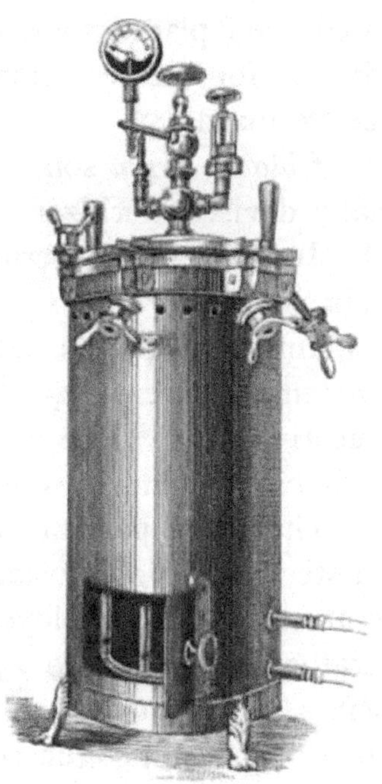

Autoclave.

to their spores, than steam in the autoclave at a pressure of an additional one-half to one atmosphere; the steam then reaches a temperature of about 110° to 120°C. Under these conditions culture-media may be sufficiently sterilized in the autoclave in fifteen minutes, and at a single sterilization. The autoclave consists of a metal cylinder with a movable top, which is fastened down tightly during sterilization. It is furnished with a thermometer, a pressure-gauge, and a safety-valve which allows the steam to escape if too high a pressure is attained. Heat is furnished by a gas-burner underneath. The lower part of the cylinder contains water. The objects to be sterilized are supported above this water on a perforated bottom or shelf.

It is necessary to follow certain precautions in the use of the autoclave, especially during cooling. The apparatus must not be opened while the steam contained within it is still under pressure, as there may be a sudden evolution of steam upon the removal of the pressure which may blow the media out of their tubes and flasks. The apparatus must, therefore, be kept closed until the gauge shows that the atmospheric pressure is as great as the pressure within, or, what is equivalent, until the temperature has fallen to 100° C. Gelatin, especially, may be damaged by sterilization with the autoclave, if it be heated too long or to too high a temperature.

FIG. 16.

Kitasato Filter.

Sterilization by Filtration.—Ordinary filters are useless for this purpose, but the tubes or bougies of unglazed porcelain devised by Pasteur and Chamberland are effective when properly employed. The Berkenfeld filter employs bougies made of infusorial earth, and its pores are larger than those of the Pasteur filter. Both of these are made in several grades according to the coarseness or fineness of the pores. The coarser of these filters permit the passage of very small bacteria. Bacteria of average size, like bacillus coli communis, may grow through the pores in the walls of both the Berkenfeld and Pasteur filters if sufficient nutrient material is present to permit of their multiplication.[1]

Filters of these kinds are widely used for water, and will be spoken of in connection with the chapter on water. Similar tubes are employed for the filtration of certain organic nutrient media whose ingredients would be damaged by sterilization with heat, chiefly extracts of organs, such as the thymus gland. The soluble " toxins " of bacteria may be obtained by filtration of fluid-cultures through such tubes, which remove the bacteria (Fig. 16). These fluids usually filter very slowly, and filtration will have to be assisted by some form of vacuum-pump; usually the filter-pump, which is used in connection with a stream of running water, is employed. Compressed air or carbonic acid may be used to assist in forcing fluids through the filter. The filter bougies, the flasks and all parts of the apparatus must, of course, be sterilized by heat before and after using.

[1] Wherry, *Journal of Medical Research,* Vol. VIII., 1902.

CHAPTER III.

CULTURE-MEDIA.

CULTURE-MEDIA are substances in which bacteria are artificially cultivated. The number of such substances is very large, different materials being suited to different purposes and to different kinds of bacteria. The most important ones are nutrient bouillon or beef-tea, nutrient gelatin, and nutrient agar-agar. The two last have a jelly-like consistency, owing to the addition of a gelatinizing substance, but otherwise are of the same composition as bouillon.

Nutrient Bouillon.

Beef-extract (such as Liebig's)................	3 grams.
Peptone, pure (Witte's)[1].....................	10 grams.
Sodium chloride (common salt)...............	5 grams.
Water	1 liter.

The solid ingredients are dissolved in water, and the mixture is boiled for a few minutes. It is made neutral or very faintly alkaline by the addition of a solution of sodium hydroxide, drop by drop, the reaction being tested at intervals with litmus-paper. The bouillon may now be filtered through filter-paper. The filter-paper should be folded and creased as is done by pharmacists; it is usually placed in a glass funnel, and should be moistened with water before using. After filtration the medium is to be placed in properly plugged tubes or flasks, and is to be sterilized once in the autoclave, or in the steam sterilizer for fifteen minutes

[1] Commercial "peptones" are mixtures of albumose and a small amount of peptone.

or longer on each of three consecutive days. When precipitates form, they are usually caused by a too alkaline reaction. That may be corrected by the addition of a little weak hydrochloric acid, drop by drop, testing frequently with litmus-paper.

A more accurate way of obtaining the proper reaction is Schultz's method. Take of the bouillon 10 c.c.; add a few drops of phenolphthalein[1] (alcoholic solution ½ per cent.) ; with a burette add, drop by drop, a solution of caustic soda 0.4 per cent. until a faint red color appears, which indicates the beginning of the alkaline reaction. This procedure is followed with three samples. The amount of soda solution required in each case is noted and the average taken. If now, on the average, for each 10 c.c. of bouillon 1 c.c. of soda solution needs to be added, for 1,000 c.c. of bouillon 100 c.c. of the soda solution must be added; only, instead of adding a weak soda solution, one-tenth as much is taken of a solution ten times as strong.

Another method of making bouillon is to use, instead of beef-extract, 500 grams (one pound) of finely chopped, lean beef, which is placed in one liter of water and kept on ice for twenty-four hours. It is strained, thoroughly cooked to coagulate the albumen in it, filtered, and a liter of fluid obtained, adding water if necessary. The peptone and salt are then added and the medium heated to dissolve them. It is then neutralized, filtered, and sterilized. Although bouillon made with solid beef-extract is convenient and serviceable for most purposes, it is advisable to use fresh meat when the bouillon is to be employed for the development of bacterial toxins. Fresh meat should also be used in the preparation of either bouillon, gelatin or agar-agar when new species of bacteria are being studied for publication.

[1] In neutralizing an acid culture-medium it has been found that when the medium appears to be neutral or slightly alkaline to litmus, it may still be acid if phenolphthalein be employed as an indicator. Fuller, *Journal American Public Health Association.* 1895.

In both of these cases the recommendations of the American Public Health Association should be followed.[1]

These also advise that media be neutralized by *titration.*—

The following solutions are required: $\frac{1}{2}$ per cent. phenolphthalein in 50 per cent. alcohol, normal[2] $\left(\frac{N}{I}\right)$ and twentieth normal $\left(\frac{N}{20}\right)$ solutions of sodium hydroxide and of hydrochloric acid.

To 5 c.c. of bouillon in a porcelain evaporating dish add 45 c.c. of distilled water; boil three minutes; add 1 c.c. of phenolphthalein solution, and proceed with the titration while still hot. As the reaction will usually be found acid, add from a burette $\frac{N}{20}$ sodium hydroxide solution, stirring constantly, until a decided pink color develops in the entire solution. The color reaction indicates the more or less arbitrarily adopted neutral point. Repeat this procedure with three different portions of bouillon, and determine the average amount of $\frac{N}{20}$ sodium hydroxide required. It is now possible to calculate the amount

[1] See the Report of the Committee of the American Public Health Association entitled Procedures Recommended for the Study of Bacteria. 1898. Rumford Press, Concord, N. H.

[2] A normal solution of any substance contains, in a liter, as many grams of the substance as there are units in its molecular weight, in case it contains a single atom of replaceable hydrogen. If it has two atoms of replaceable hydrogen the number of grams used equals the molecular weight divided by two; and so on. Thus the molecular weight of sodium hydroxide is 40, and its normal solution contains 40 grams of sodium hydroxide in a liter. It is not expedient to prepare normal solutions of sodium hydroxide by weight. For convenience, crystallized oxalic acid is used as a starting point in making normal solutions. Its molecular weight, including a molecule of water of crystallization, is 123. As it is a dibasic acid (having two atoms of replaceable hydrogen), half of this weight, or 62.5 grams, per liter, is taken. Any $\frac{N}{I}$ acid solution will exactly neutralize an equal volume of any $\frac{N}{I}$ alkaline solution. To make $\frac{N}{I}$ sodium hydroxide solution, add about 41 grams of pure caustic soda to a liter of distilled water. Find the amount of this solution needed to exactly neutralize 1 c.c. of $\frac{N}{I}$ solution of oxalic acid; this amount contains the quantity of sodium hydroxide which should be present in 1 c.c. of a normal solution. It is now possible to calculate the amount of distilled water to be added in order that 1 c.c. of the sodium hydroxide solution may neutralize 1 c.c. of the $\frac{N}{I}$ solution of oxalic acid. With an $\frac{N}{I}$ solution of sodium hydroxide as a standard, an $\frac{N}{I}$ solution of hydrochloric acid may be prepared. Twentieth normal solutions have one-twentieth the strength of normal solutions.

of $\frac{N}{1}$ sodium hydroxide needed to neutralize the whole quantity of bouillon. This should be added. The bouillon should then be boiled for ten minutes, and again titrated. It will usually be found acid. The deficiency should be corrected by adding the necessary amount of $\frac{N}{1}$ sodium hydroxide. It should be boiled again, and again titrated, and any deficiency made good. It is rarely necessary to repeat the process, except to determine that the neutral point has been reached. After neutralizing it is boiled thirty minutes and filtered. Enough $\frac{N}{1}$ hydrochloric acid or sodium hydroxide is added to give the degree of acidity or alkalinity desired. It is then sterilized.

An acid reaction may be denoted by $+$, an alkaline by $-$. The degree of acidity or alkalinity may be indicated by the amount of $\frac{N}{1}$ solution required to render the medium neutral to phenolphthalein, thus $+$ 1.5 signifies that a medium is acid, and requires 1.5 per cent. of $\frac{N}{1}$ sodium hydroxide to neutralize it.

A reaction of $+$ 1.5 is recommended as the optimum. There is much disagreement as to what reaction is most favorable for the growth of the majority of species of bacteria. In any case the degree of reaction should be noted in descriptions.

Bouillon may be modified by the addition to it of other substances, the most important of which are glycerine (6 per cent.) and sugars,—as dextrose,[1] saccharose or lactose (1 per cent.). It is better to sterilize media containing sugars in the steam sterilizer by the fractional method than in the autoclave, where decomposition of the sugars may occur.

Dextrose-free Bouillon.—Ordinary bouillon often contains some muscle-sugar, which is objectionable if fermentation tests with lactose or saccharose are to be made. To secure bouillon free of sugar, beef-infusion is prepared from fresh meat, and is inoculated in the evening with a quantity of bacillus coli communis, and kept in the incubator. Early next morning it is boiled, filtered, peptone and salt added, and the bouillon is prepared as usual.[2]

Nutrient Gelatin.

Beef-extract	3 grams.
Peptone	10 grams.
Sodium chloride	5 grams.
Gelatin (best gold label)	100 grams.
Water	1 liter.

[1] Dextrose is the principal ingredient of commercial grape-sugar and should be obtained in a pure condition.

[2] See T. Smith, *Journal of Experimental Medicine*, Vol. II., p. 546.

Dissolve the ingredients in the water, stirring actively to prevent burning at the bottom. It is best to conduct the operations in granite- or enamel-ware vessels over a large Bunsen or rose-burner. Neutralize with sodium hydroxide solution (see page 71). The reaction at the beginning will usually be found to be quite acid. Allow the mixture to cool until below 60° C., and add the whites of one or two eggs which have been beaten up with a little water; stir in thoroughly. Heat the mixture to the boiling-point; stir at the bottom to prevent burning and at the same time avoid as far as possible breaking the coagulum of egg-albumen which forms at the surface. Boil for ten minutes. Filter while hot. The filtration may be done through folded filter-paper which has been moistened. It is well to fasten a piece of coarse cheese-cloth over the top of the funnel to catch the large particles of coagulated albumen. Place in suitable tubes or flasks plugged with cotton, and sterilize once in the autoclave, or, preferably, in the steam sterilizer for fifteen minutes on each of three consecutive days. Gelatin is injured by too prolonged boiling and loses its solidifying qualities. Neutralization may be with litmus paper or by titration. The remarks on pages 72 to 74 with regard to the use of fresh beef and the titration method for the preparation of bouillon apply equally to gelatin.

Instead of filter-paper, some prefer to filter through several layers of absorbent cotton placed inside of the moistened glass funnel, the top of which is covered with coarse cheese-cloth. This expedient answers very well.

If the product appears cloudy after it has been sterilized, it may be that the egg-albumen was incompletely coagulated in the first place or that the reaction has been made too alkaline. In any case it will be desirable to melt it and filter a second time, correcting the reaction with hydrochloric acid if necessary. It may be well to stir in another egg to entangle the opaque particles; then to boil a second time and filter.

The medium is sometimes modified by adding to it other substances, as sugar, glycerin, etc. The solidifying property of the gelatin must be carefully guarded, and too much boiling is to be avoided. Certain bacteria, it will be found, have the property of causing gelatin to become fluid. Gelatin melts at about 25° C. and solidifies at about 10° C. It cannot be used in the incubator, where it would liquefy at the temperature of 38° C. In hot weather it may be necessary to use 150 grams of dry gelatin to the liter. Nutrient gelatin is usually spoken of simply as " gelatin."

Nutrient Agar-agar.—*Agar-agar* (French, *gélose*) is a kind of vegetable gelatin which comes from the southern and eastern coast of Asia. It melts with much greater difficulty than gelatin.

The medium is not quite transparent. The finished medium is commonly called " agar."

Beef-extract	3 grams.
Peptone	10 grams.
Sodium chloride........................	5 grams.
Agar	10 grams.
Water	1 liter.

The dry agar, cut fine, is to be dissolved in water over a flame. It should be boiled for from one-half hour to two hours, skimming off the scum which forms on the surface from time to time. The beef-extract, peptone and sodium chloride are dissolved in a liter of water, boiled and neutralized. Add the agar now in solution in a small quantity of water. The reaction of the agar alone is faintly alkaline. Mix thoroughly; the bulk of the mixture is a little more than a liter, and should be reduced to a liter after the subsequent boiling. Cool to about 60° C.; stir in the whites of one or two eggs and boil thoroughly. Avoid breaking the coagulum of egg which is designed to entangle the solid particles that make the medium cloudy; stir at the bottom, however, to prevent burning. Filter while hot, using filter-paper or absorbent cotton covered with cheese-cloth. The hot water funnel originally devised for the filtration of agar is not necessary. If filtration is slow, the funnel and flask may be placed inside of the steam sterilizer and kept heated during filtration. The medium is collected in suitable flasks or tubes plugged with cotton, and sterilized once in the autoclave or in the ordinary steam sterilizer for fifteen minutes on each of three consecutive days. As agar is frequently used for smear-cultures where a slanted medium is desired, some of the tubes may be allowed to cool in a slanting position. It is not well to keep on hand many tubes which have been slanted, as the medium dries more rapidly. Agar is not liquefied by bacteria as is gelatin. Its solidifying qualities are impaired somewhat if the reaction be acid.

The remarks on pages 72 to 74 with regard to the use of fresh beef and the titration method for the preparation of bouillon apply equally to agar-agar.

Glycerin-agar is used extensively. It is agar, made as above directed, to which 6 per cent. of glycerin is added before sterilization. It is very useful in cultivating the bacilli of tuberculosis and diphtheria.

Sugar-agar.—Before sterilizing, 1 per cent. of either dextrose, lactose, saccharose, or other sugars may be added to agar. With media containing sugar, litmus forms a useful indicator of the production of acid. Enough tincture of litmus is used to give the medium a blue color before sterilization; the litmus is somewhat unstable and prone to change its color during sterilization. Neutral red may also be added in the same manner; its color is said to be changed by certain bacteria and not by others (see bacillus of typhoid fever, and bacillus coli communis, Part IV.).

Potato.—The potatoes are washed, a slice is removed from each end, and with an apple-corer or cork-borer a cylinder is cut out. This cylinder is divided diagonally into two pieces. The pieces are washed in running water for twelve to eighteen hours. They are placed in test-tubes containing a little water to keep the potato moist, and are supported from the bottom on a piece of glass tubing about 1 to 2 cm. in length (or on cotton, or in a specially devised form of tube with a constriction at the bottom). The tubes are plugged, and sterilized as with other media. Sterilization, however, must be thorough on account of the danger of contamination with the extremely resistant spores of the potato bacillus. Potato is best when freshly prepared; it is likely to become dry and discolored with keeping. It is a very useful medium; certain growths on it, like those of the bacillus of typhoid fever or of glanders, and those of chromogenic bacteria, are very characteristic.

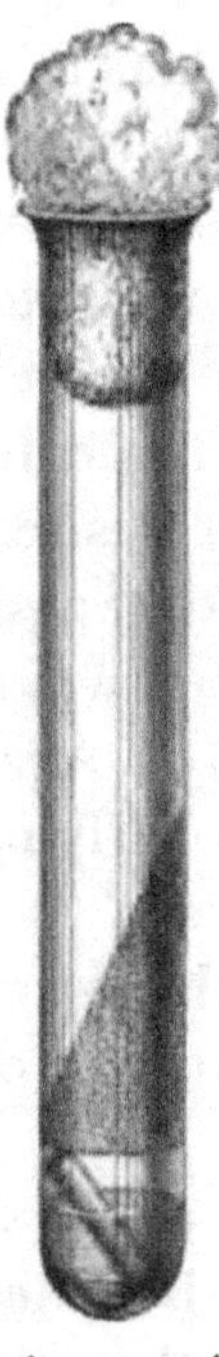

FIG. 17.

Tube containing Potato.

Milk.—Milk fresh as possible is placed in a covered jar, sterilized for fifteen minutes, and then kept on ice for twenty-four hours. At the end of that time the middle portion is removed by means of a siphon. The upper and lower layers must not be taken; the upper part contains cream, and the lower part particles of dirt, both of which are to be avoided. About 7 to 10 c.c. are to be run into each test-tube. The tube is plugged with cotton, and sterilized as usual. When milk is contaminated with spores of the hay or potato bacillus it is sometimes very difficult to sterilize, a fact of much importance in connection with the feeding of children, where the fractional method of sterilization and the use of the autoclave are impracticable.

The coagulation of milk, which is accomplished by certain bacteria, is a very valuable differential point. A little litmus tincture may be added to the tubes of milk before sterilizing, until they acquire a blue color, to indicate whether or not acids are formed by the bacteria which are afterwards cultivated in the milk.

Dunham's Peptone Solution.

Peptone .... 10 grams.
Sodium chloride............................. 5 grams.
Water .... 1 liter.

Boil, filter, sterilize in the usual manner.

Dunham's solution is valuable to test the development of indol by bacteria (see Part II., Chapter II.). The development of acids may be detected after the addition of 2 per cent. of rosolic acid solution (.5 per cent. solution in alcohol); alkaline solutions give a clear rose-color which disappears in the presence of acids.

Blood-serum.—The blood of the ox or cow may be obtained easily at the abattoir. It should be collected in a clean jar. When it has coagulated, the clot should be separated from the sides of the jar with a glass rod. It may be left on the ice for from twenty-four to forty-eight hours. At the end of that time the serum will have separated from the clot and may be drawn off with a siphon into tubes. These tubes are sterilized for the first time in a slanting position as the first sterilization coagulates the serum. The coagulation may be done advantageously, as advised by Councilman and Mallory, in the hot-air sterilizer at a temperature below the boiling-point. After coagulation, sterilize as usual. This serum makes an opaque medium of a cream color. Blood-serum may be sterilized in the special form of sterilizer devised for it. A clear blood-serum is to be obtained by sterilization at a temperature of 58° C. for one hour, on each of six days, if a fluid medium is desired, or of 75° C. on each of four days if the serum is to be solidified. In the latter case the tubes are to be placed in an inclined position. (See page 67.) Opaque, coagulated blood-serum has most of the advantages of the clear medium. Blood-serum may be secured from small animals by collecting blood directly from the vessels, using very great care to obtain the blood in a sterile condition; and the serum may be separated and stored in a fluid state. Human blood-serum is sometimes obtained from the placental blood, sometimes from serous pleural transudates or from hydrocele fluid. The preservation of blood-serum is sometimes accomplished with chloroform, of which 1 per cent. is to be added to the medium; in this manner the serum may be preserved for a long time. It may be divided into tubes, solidified and sterilized as required; the chloroform will be driven off by the heat, owing to its volatility. Blood-serum media which are sterilized at low temperatures should be tested for twenty-four hours in the incubator to prove that sterilization has been effective; if it has not, development of the contaminating bacteria will take place and be visible to the eye.

It will be impossible to do more than merely mention some of the most important of the other culture-media.

Löffler's blood-serum consists of one part of bouillon containing 1 per cent. of glucose, and three parts of blood-serum. It is sterilized like ordinary blood-serum. It is used largely for the cultivation of the bacillus of diphtheria.

Blood-serum-agar is a medium made with considerable difficulty, but very valuable for the cultivation of the gonococcus. One part of placental blood-serum, or pleuritic serum, or hydrocele fluid, is mixed with one to two parts of nutrient agar in the fluid condition. It must be divided into tubes before solidification. Solidify in a slanting position; subsequently sterilize at 75° C. so as not to coagulate the albumen of the blood-serum. The nutrient agar in this case should contain 2 per cent. of dry agar. Another expedient has also been to smear a little blood over the surface of a tube of nutrient agar—*blood-agar*—used for cultivating the bacillus of influenza. *Marmorek's blood-serum* is supposed to assist in maintaining the very evanescent virulence of the streptococci; it consists of bouillon mixed with human blood-serum, ass's serum or horse's serum.

Guarnieri's medium consists of a mixture of gelatin and agar.

Media containing *fat* were employed by Sommaruga to test the ability of bacteria to decompose fats. Clarified beef-suet or olive-oil in the proportion of 1 or 2 per cent. is added to gelatin or agar. The fat must be mixed with the melted medium; it is to be shaken and then rapidly cooled in a freezing-mixture after the last sterilization.

Fresh eggs in their shells may be used without other preparation than washing the surface thoroughly with bichloride of mercury solution; or after sterilization by steam, which of course coagulates the albumen. The egg is easily inoculated through a small opening made with a heated needle, which may be closed afterward with collodion. Hueppe recommended eggs closed in this manner for the cultivation of anaërobic bacteria. Egg-albumen has been used as a constituent of various media. Dorset[1] states that good results may be secured when eggs are used as a culture-medium for tubercle bacilli. The yolk and the white are mixed, poured into tubes, slanted, coagulated, and sterilized. Just before using pour into the tube a few drops of sterile distilled water to moisten the medium.

Bread-paste (finely-divided dry bread, mixed with water and sterilized) is used for the cultivation of moulds. *Sa-*

[1] *American Medicine*, April 5, 1902.

bouraud recommends the following for the cultivation of the trichophyton fungus:

Peptone	5 grams.
Maltose	3.8 grams.
Agar	1.3 grams.
Water	100 c.c.

Test-tubes.—Bacteria are generally cultivated in test-tubes. A convenient size is one ⅝ of an inch in diameter

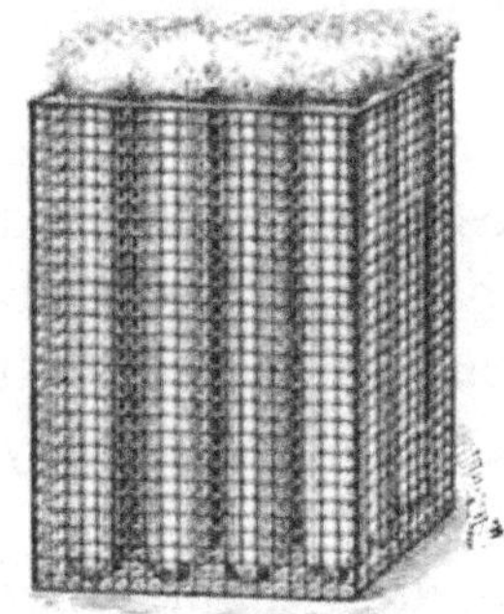

FIG. 18.

Wire Basket for Test-tubes.

and 5 inches in length. The tubes should be of a heavier glass than in those used for ordinary chemical work. The New York Board of Health, and some others, use a tube three inches in length without a flange for the cultivation of the diphtheria bacillus on Löffler's blood-serum mixture. Test-tubes should be thoroughly cleaned with a swab before using; they should be boiled with washing-soda, rinsed, filled with hydrochloric acid solution, rinsed, and inverted to drain away the fluid.

Plugs of raw cotton or cotton batting are employed as stoppers. Some prefer absorbent cotton, but it is likely to become soggy after exposure to steam. The plug should fit smoothly; creases and cracks around the edges are to be avoided. The plug should be tight enough to sustain the weight of the tube when held by the plug. These plugs prevent bacteria from entering or leaving the tubes.

Sterilization of Test-tubes.—The tubes are to be sterilized in a hot-air sterilizer for one hour, at a temperature of 150° C. The cotton should acquire a light brown color but should not be burned. If the plugs touch the sides of the sterilizer or lie against the bottom they may be scorched.

The necessity for sterilization of the tubes before filling

them with the medium has been questioned, and it is probably unnecessary as far as the preservation of the culture-medium is concerned, but it will be found that the cotton plugs fit much better after sterilization with dry heat. During this and subsequent sterilizations the tubes are held in a wire basket.

Filling of the Tubes.—A special funnel closed with a stop-cock for filling tubes with liquefied media is often recommended. They may readily be filled with an ordinary funnel of small size. During the filling, the neck of the test-tube where it comes in contact with the cotton *must not be wet* with the medium. Ordinarily about 7 to 10 c.c. are placed in a test-tube. For Esmarch's roll-tubes a somewhat smaller quantity is desirable.

The sterilization of tubes containing culture-media is always done by steam, and has been sufficiently described. It is to be remembered that the solidifying power of gelatin is impaired by too prolonged heating, while heating is less likely to damage other culture-media. The media which are sterilized at a low temperature (70° C.) should be tested for two days in the incubator to determine whether sterilization has been effective. It is the universal experience in bacteriological laboratories that occasionally culture-media will become contaminated with extremely resistant spores which fail to be sterilized by the ordinary processes, an occurrence which causes great annoyance and calls for the exercise of much patience. Sometimes, also, moulds attach themselves to the plugs, especially if they are moist, and send their filaments down through the cotton; finally, having reached the lower edge of the cotton, their spores may fall upon the medium, grow there and ruin it.

8

CHAPTER IV.

THE CULTIVATION OF BACTERIA.

Inoculation of the Tubes.—The air of the laboratory should be as quiet as possible, to lessen the chances of contamination by bacteria clinging to particles of dust. Avoid working where there may be draughts or gusts of air or near an open window. Spores are blown from the surfaces of moulds, like thistle-down, and are constantly being wafted about in the air. Given any material containing bacteria, for example a pure culture of some well-known species, a very minute portion is to be introduced into a tube containing the sterile culture-medium. The introduction is effected with a straight platinum wire, or with a platinum wire loop. The platinum is to be heated red-hot before using, and then allowed to cool. It is also to be heated red-hot after using. The plug of the test-tube is to be withdrawn, twisting it slightly, taking it between the third and fourth fingers of the left hand, with the part that projects into the tube pointing toward the back of the hand. It must not be allowed to touch any object while the inoculation is going on. Pass the neck of the tube through the flame. If any of the cotton adheres to the neck of the tube, pull the cotton away with sterilized forceps, while the neck of the tube touches the flame, so that the threads of cotton may be burned and not fly into the air of the room. The tube is held as nearly horizontal as possible. The tube is to be held in the left hand between the thumb and forefinger, the tube resting upon the palm, and the neck of the tube pointing upward and to the right. When two tubes are being used at the same time, as is often necessary, they are placed side by side between the thumb and forefinger of the left hand. The two plugs are held between the second and third and the third and fourth fingers of the left hand, respectively. The wire may now be passed into the first tube, which we will suppose to hold some material containing bacteria, and a little of this material may be removed on the tip of the wire from the first tube to the second. When the needle is introduced into or removed from either tube it should not touch the side of the tube at any point, and should only come in contact with the region desired. After inoculation of the second tube has been effected the wire is to be heated to a red heat in the flame, the necks of the tubes are to be passed through the flame, and the plugs are to be returned to their

FIG. 19.

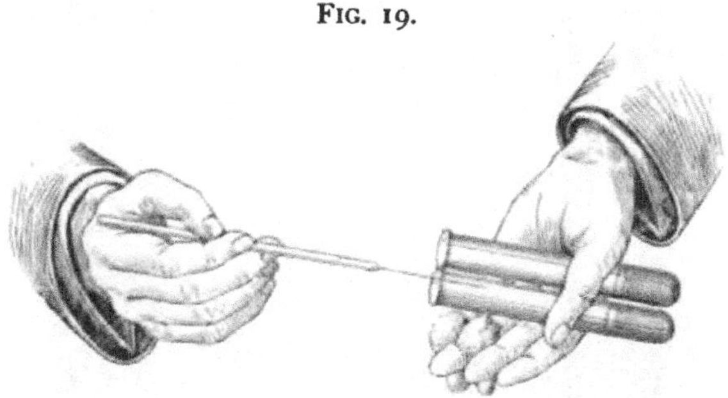

Manner of Holding Tubes.

respective tubes. When the wet wire is to be sterilized in the flame it should be approached to the flame gradually, so as to dry the material on it before burning it, in order to avoid "sputtering" (see page 33). It is well from the start to train one's self to *sterilize the platinum wire every time it is taken from the table and before it is laid down again.* The platinum wire loop may be used in the same manner as the straight wire, especially when a substance containing a small number of bacteria is being handled.

When a tube of gelatin is to be inoculated the wire is usually introduced into the medium vertically, " stab-culture "; when a medium with a slanted surface is employed, as agar, potato or blood-serum, the needle should lightly streak the surface, "smear-culture" (Figs. 20 and 21).

The safety and success of this method of inoculation depend upon a principle which has been established by long and repeated observation, namely, that bacteria do not of themselves leave a moist surface. They should not, there-

FIG. 20.

FIG. 21.

Stab-Culture.
A rubber stopper may be
used to prevent drying,
see page 91.

Smear-Culture.
This tube shows the rubber
cap used to prevent
drying.

fore, rise from the surface of the moist culture-medium, nor drop from the needle during its transit, if proper care be exercised. They may be thrown into the air if the needle be allowed to sputter in the flame.

If, by any accident, drops of infectious material should fall upon a surface like the table, they should be covered at once with bichloride of mercury solution 1–1000. A good way is to cover the spot with a piece of blotting-paper wet with the solution; place a bell-jar over it and leave for several hours. If infectious material should reach the hands or clothing, they should be thoroughly soaked in the bichloride solution. When working with pathogenic bacteria it is well to wash the hands in this solution and with soap and water, as a routine procedure, before leaving the laboratory.

To maintain their vitality bacteria need to be transplanted from one tube to another occasionally; the time varies greatly with different species. Many bacteria grow on culture-media with difficulty at the first inoculation, but having become accustomed· to their artificial surroundings, as it were, they may be propagated easily afterward; this is especially true of the bacillus tuberculosis.

Some bacteria flourish better on one culture-medium than another. The bacillus tuberculosis grows best on blood-serum and glycerin-agar; the bacillus of diphtheria grows best on Löffler's blood-serum; the gonococcus on human serum-agar.

The virulence of most pathogenic bacteria becomes diminished after prolonged cultivation upon media. Sometimes the virulence is lost very quickly, for example, the streptococcus pyogenes and micrococcus lanceolatus of pneumonia.

Incubators.—Many bacteria flourish best at a temperature about that of the human body, 38° C. Some species will grow only at this temperature. The pathogenic bacteria in particular, for the most part, thrive best at a point near the body temperature.

The incubator is a box made of copper, having double

walls, the space between the two being filled with water. The outer surface is covered with some non-conductor of heat, such as felt or asbestos. At one side is a door, which

FIG. 22.

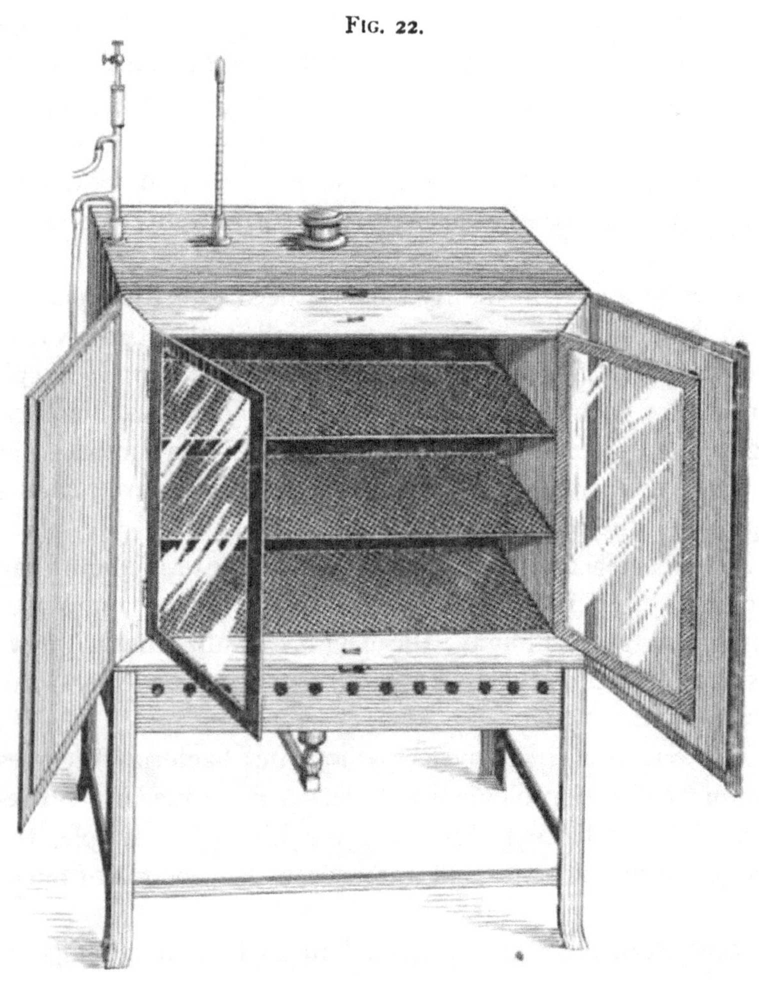

Incubator.

is also double. The inner door is of glass, the outer door is of copper covered with asbestos. At one side is a gauge which indicates the level at which the water stands in the water-jacket. The roof is perforated with several holes, some of which permit the circulation of the air in the air-chamber inside the box; some of them enter the water-jacket. A thermometer passes through one of these holes into the interior of the air-chamber, and often another into the water standing in the water-jacket. A gas-regulator passes through another hole, and is immersed in the water

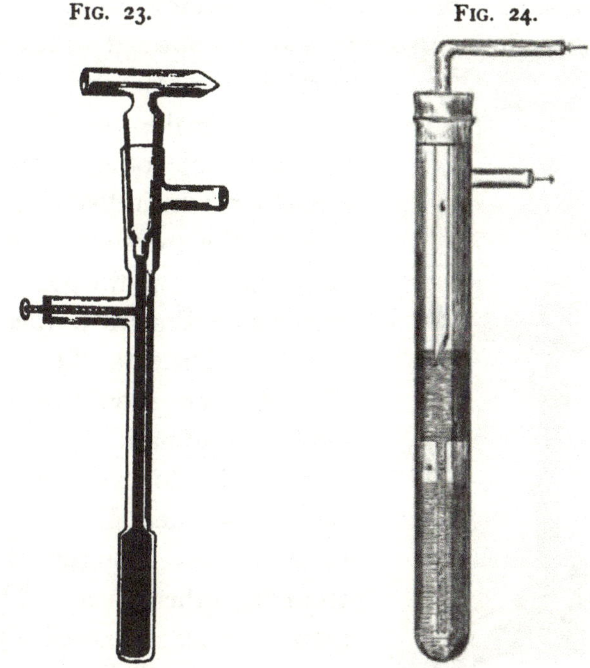

FIG. 23.

FIG. 24.

Reichert's Gas-regulator.

Mercurial Gas-regulator. *a.* Chamber containing volatile hydrocarbon. *b.* Capillary opening.

standing in the water-jacket. There are various forms of gas-regulators more or less complicated. In general they consist usually of a tube containing mercury; into this tube are two openings, one for the entrance and the other for the exit of gas. The gas enters through a small tube, which is cut off diagonally at the bottom, and which projects into the surface of the mercury. Heating the water in the water-jacket causes expansion of the mercury, which rises, and, little by little, cuts off the inflow of gas through this tube. The flow is never completely cut off, as there is a capillary opening in the tube considerably above any point to which the mercury could possibly rise, which will always allow the flow of a small quantity of gas (Fig. 24, *b*). This diagram also shows a modification of the simple form of regulator, in the shape of a partition which divides off a lower chamber, which contains mercury and is connected with the upper

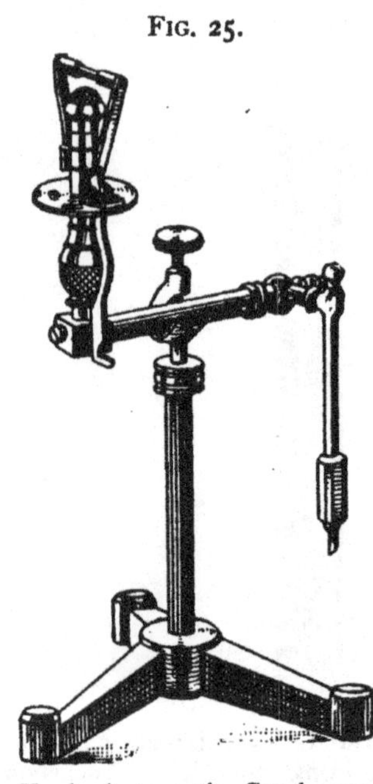

FIG. 25.

Koch Automatic Gas-burner.

part by a glass tube. The purpose is to make use of the elastic properties of some volatile fluid, like ether, which floats on the surface of the mercury at *a*. The gas coming from the gas-regulator passes to a Bunsen burner, which stands underneath the incubator. This burner should have some kind of automatic device for cutting off the flow of gas in case it becomes accidentally extinguished by a sudden draught of air or from any other cause. The automatic burner invented by Koch is an ingenious, simple and effective device. A bar of metal stands above the flame; by its expansion, through a system of levers, it supports a weight; the weight controls a gas-cock. While the flame is burning the expansion of the metal holds the weight horizontally; if the flame becomes extinguished, the metal contracts, the weight falls, and cuts off the flow of gas. Some inconvenience will arise from irregularities in the flow of gas from the main supply-pipe. Any incubator will vary a little from such causes. In the experience of the writer, natural gas is of such variable pressure as to be entirely useless. Fluctuations of the temperature within the incubator depend very largely upon the external temperature. Therefore the incubator should, as far as is practicable, be protected from sudden draughts of cold air and should be kept in a room having as equable a temperature as possible.

Culture-tubes which are being kept in the incubator are likely to become dry if their stay is prolonged. In such cases they should be covered with rubber caps, tin-foil, sealing-wax, paraffin, or some other device to prevent evaporation. If rubber caps are used, they should be left in 1–1000 bichloride of mercury solution for an hour, and the cotton plugs should be singed in the flame, before putting them on. (Fig. 21.) The writer prefers rubber stoppers, which may be boiled and stored in bichloride of mercury solution. Cut the cotton plug even with the edge of the tube; singe it in the flame; push it into the tube about 1 cm.; and insert the rubber stopper. (Fig. 20.)

CULTIVATION OF ANAËROBIC BACTERIA.

The cultivation of anaërobic bacteria is done best in a medium containing 1 to 2 per cent. of dextrose. The tube should contain a large quantity of the culture-medium. Just before using, the medium should be boiled for a few minutes. Inoculate the tube after cooling, but while the medium is fluid. Anaërobes may be cultivated in the closed arm of the fermentation-tube (see Fig. 46), but the opening between the two arms of the tube must be small.

Buchner's method for the cultivation of anaërobes: Into a bottle or tube which can be tightly stoppered, pour 10 c.c.

of a 6 per cent. solution of sodium or potassium hydroxide, for each 100 c.c. of air contained in the jar. Add one gram of pyrogallic acid for each 10 c.c. of solution. The culture-tube is placed inside of the larger bottle or tube, supported above the bottom, and the stopper, smeared with paraffin, is inserted. The mixture of pyrogallic acid and potassium hydroxide possesses the property of absorbing oxygen.

Wright's Modification of Buchner's method: The tube of culture-medium is to be plugged with *absorbent cotton,* using a plug of large size. The culture-medium is inoculated in the usual way. The plug is cut off close to the neck of the tube, and is then pushed into the tube about 1 centimeter. Now allow a watery solution of pyrogallic acid to run into the plug, and then a watery solution of sodium or potassium hydroxide. Close quickly and tightly with a rubber stopper. Wright recommends that the first solution be freshly made and consist of about equal volumes of pyrogallic acid and water, and that the second solution contain 1 part of sodium hydroxide and 2 parts of water. With 6 inch test-tubes, $\frac{3}{4}$ inch diameter, the amounts advised are — $\frac{1}{2}$ c.c. solution of pyrogallic acid, 1 c.c. solution of sodium hydroxide.

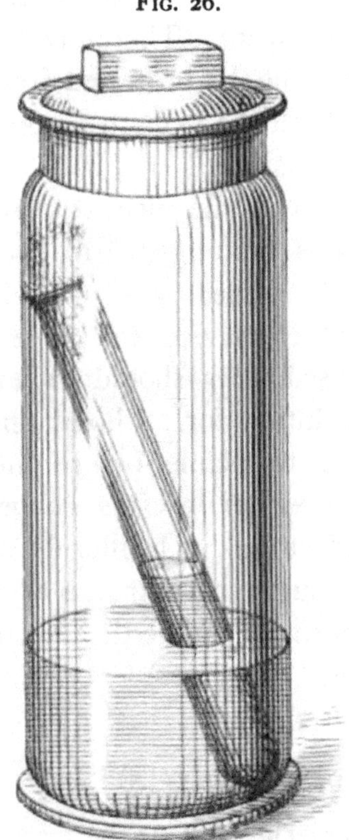

FIG. 26.

Arrangement of Tubes for Cultivation of Anaërobes by Buchner's Method.

Cultivation of Anaërobic Bacteria under Hydrogen: Method of Fränkel: A test-tube containing a large amount of the liquefied culture-medium is closed with a sterilized rubber stopper, through which pass two sterilized glass tubes, bent above the stopper at a right angle. One of these tubes is cut off just underneath the stopper, and the other is long enough to project nearly to the bottom of the culture-tube. The horizontal projecting parts are drawn to a small caliber at some point, although not quite closed, to facilitate sealing later on. Through the longer of these tubes hydrogen gas is passed until the atmosphere inside of the culture-tube is pure hydrogen, entirely free from mixture with air. The horizontal parts of the small glass tubes projecting from the stopper are then sealed in the flame at the places where they were previously drawn out to a small ·caliber, and the tubes are thus closed. (Fig. 27.)

The stopper should be surrounded with melted paraffin. A tube prepared according to this plan may, if desired, be converted into an Esmarch roll-tube. The hydrogen is generated according to the common method with *pure* zinc and *pure* sulphuric acid, 25 to 30 per cent. The precautions advised by chemists for the generation of hydrogen must be carefully followed, because when hydrogen mixed with oxygen or air is ignited a violent and disastrous explosion may occur.

The well-known Kipp's generator may be used. First

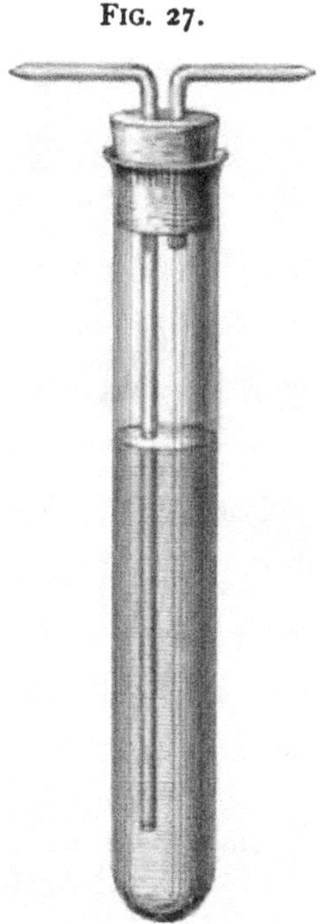

Cultivation of Anaërobes by Fränkel's Method.

let the reservoir fill with hydrogen; then allow its contents to escape. This should be repeated, after which some of the hydrogen may be collected in an inverted test-tube under water. When this sample is ignited, it should burn without any explosion; otherwise the hydrogen is not yet ready to use. The hydrogen should bubble through the medium five minutes or more.

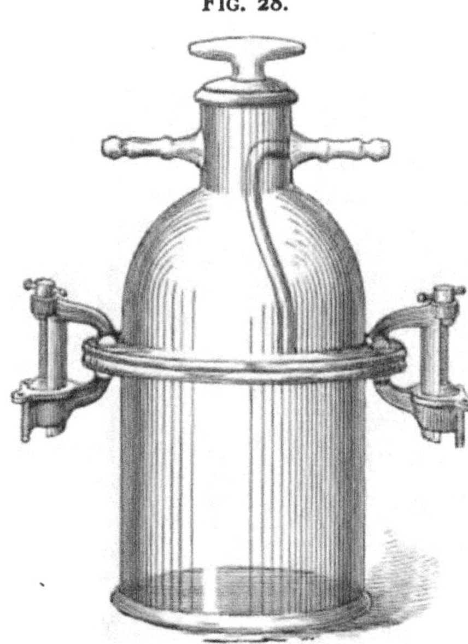

FIG. 28.

Novy's Jar for the Cultivation
of Anaërobes.

The inconvenience of sealing the tubes in the flame, as has to be done in Fränkel's and other methods for cultivation under hydrogen, is obviated in *Novy's apparatus.* The tubes or plates are placed in jars through which hydrogen may be conducted. The stopper, having been smeared previously with a soft wax, is sealed by giving it one-fourth of a turn.

There have been various other kinds of apparatus, usually complicated and expensive, devised for the growth of plate-cultures under hydrogen.

Other expedients for the cultivation of anaërobic bacteria are less effective. In cases where a very deep stab-culture is made in gelatin or agar, where the growth appears in the lower part of the tube by preference, it is supposed to be anaërobic. Koch covered part of the surface of a gelatin plate with a bit of sterilized mica or a cover-glass; bacteria which grew beneath this plate were considered to be anaërobic. Another method was to cover the surface of the gelatin in the cultu.e-tube with sterilized oil. W. H. Park has recommended a mixture of solid paraffin with 25 to 50 per cent. of fluid paraffin or albolene as a covering for the surface of anaërobic cultures. This mixture has a semi-solid consistency, and does not retract at the edges on cooling. The paraffin prevents the absorption of oxygen, except to a small extent at the edges. The method is useful for large quantities of culture material, as in flasks. Esmarch advised making roll-tubes, and after cooling them to fill them with a liquefied gelatin cooled down

FIG. 29.

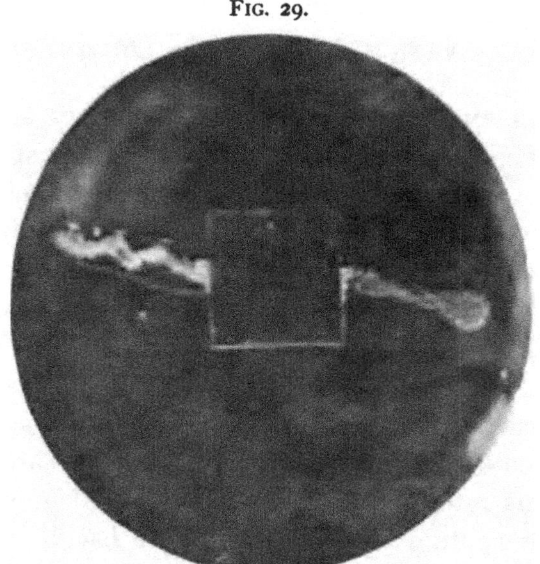

An Aërobic Organism (Potato Bacillus) which will not grow under a
cover-glass.

to near the point of solidification. Hueppe made use of eggs in their shells. The eggshell was carefully cleaned, sterilized with a solution of bichloride of mercury, washed with sterilized water and wiped dry with sterilized cotton. The end of the eggshell was punctured with a hot needle. Through the opening thus made the inoculation was accomplished. The opening was closed with collodion.

CHAPTER V.

CULTIVATION OF BACTERIA, CONTINUED.

Isolation of Bacteria.—In order to study any kind of bacteria it is necessary to have the particular species separated from other sorts with which it may be mixed. The earlier bacteriologists endeavored to separate bacteria of different sorts by successive transplantations through a series of tubes. The procedure now generally used for this purpose is the so-called plate-method of Koch. The great progress which bacteriology has made during the last twenty years is largely owing to this invention.

Pathogenic bacteria may sometimes be isolated through inoculations into animals. Thus an animal may be inoculated with sputum containing tubercle bacilli mixed with other bacteria. The animal may die of tuberculosis, and its tissues may contain tubercle bacilli in pure culture, the other bacteria having produced no important effect.

Still another method which is occasionally useful is to subject the mixture of bacteria to steam for a few minutes. If it contains very resistant spores, like those of the tetanus bacillus or hay bacillus, they may be expected to survive, and may perhaps be propagated in pure culture, everything else having been killed by the steam.

Plate-Cultures.—It is impossible in most cases to distinguish between bacteria of different varieties by microscopical examination alone. Bacteria of widely different species and quite unlike one another in their properties may present similar appearances under the microscope. The differences which they exhibit are usually apparent when they are grown in culture-media. The growth, called a *colony,* which results from the multiplication of a single bacterium, is in many cases quite characteristic for the species. By the plate-method, the individual bacteria in a mixture are separated from one another by dilution. They are fixed in place by the use of a solid medium. They are allowed to grow, and from each individual there forms a colony. It is usually possible to distinguish between *colonies* arising from different species when it was not possible to distinguish between *the individual bacteria* of these species. A convenient illustration has been suggested by Abbott. A number of seeds of different sorts may appear very much alike, and considerable difficulty may be found in distinguishing one from another with the eye. Let them be sown, however, and let plants develop from them, and these plants will easily be distinguished from one another.[1]

Method of Making Plate-cultures.—Melt *three* tubes of gelatin or agar. (There is some difficulty in keeping agar in a fluid state while dilutions are being made. It is best to have some form of water-bath with a thermometer for the purpose.) Let the liquefied tubes cool to 40° C. Take a small portion of the material to be examined—pus, for example—and introduce it with a sterilized platinum wire or loop into one of the tubes. Stir it in carefully. Remove the needle, sterilize it, and replace the plug. Mix the material introduced thoroughly with the liquefied culture-medium, taking care not to wet the plug. Now remove the plug again, and, having sterilized the platinum wire, insert it into the liquefied medium. Carry three loopfuls in

---

[1] It must be understood that *no close comparison can be drawn between higher plants,* which simply complete the development of parts potentially present in the seed, *and colonies of bacteria,* which are aggregates of individuals, the progeny of one individual of the same kind.

succession from this tube, which is No. 1, into tube No. 2; sterilize the needle; replace the plugs; mix thoroughly, without wetting the plug. Carry three loopfuls from tube No. 2 into tube No. 3 in the same manner. The original material will obviously be diluted in tube No. 1, more in tube No. 2, and still more in tube No. 3. The most convenient form of plate is that known as a Petri dish, a small glass dish about 8 cm. in diameter and 1.5 cm. in height, provided with a cover which is a little larger but of the same

FIG. 30.

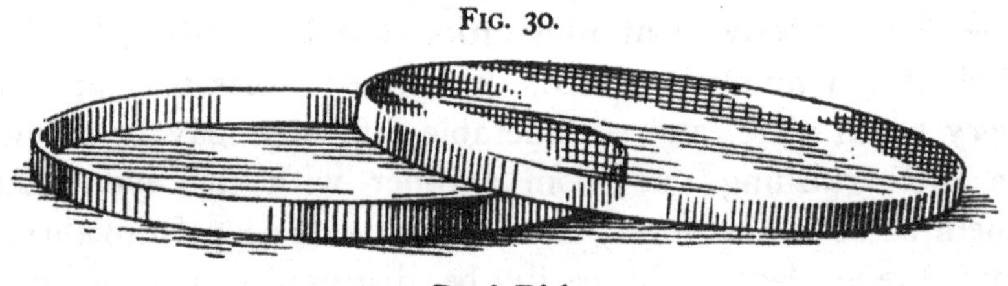

Petri Dish.

form. This dish should be cleaned and sterilized for an hour in a hot-air sterilizer at 150° C. or higher. When it is cool it may be used.

Such dishes having previously been prepared, the contents of tube No. 1 are poured into one dish, and those of tube No. 2 into another, and those of tube No. 3 into a third. They are to be labeled Nos. 1, 2, and 3.[1] In pouring proceed as follows: remove the plug of tube No. 1; heat the neck of the tube in the flame; allow it to cool, holding it in a nearly horizontal position. When the tube has cooled, lift the cover of the Petri dish a little, holding it over the dish; pour the contents of tube No. 1 into the dish, and replace the cover of the dish. The interior of the dish should be exposed as little and as short a time as possible.

[1] The labels should be moistened with the finger, which has been dipped in water. They should not be licked with the tongue. While working in the bacteriological laboratory it is best to make it a rule that no object is to be put in the mouth.

Tubes Nos. 2 and 3 are to be treated in the same manner. Burn the plugs, and fill the empty tubes with 5 per cent. solution of carbolic acid. They should be sterilized for an hour in the steam sterilizer on each of three days.

The culture-medium in the Petri dish will soon solidify. Colonies develop usually in from one to two days. In plate No. 1 they will be very numerous, in plate No. 2 less numerous, and in plate No. 3 still less numerous. Where the number is small the colonies will be widely separated and can readily be studied. They may be examined with a hand-lens, or the entire dish may be placed on the stage of the microscope and the colonies be inspected with the low power. The iris diaphragm should be partly closed and the concave mirror should be used. Dilution-cultures prepared as described in the next paragraph, where the principle is the same, are shown in Fig. 31. In tube No. 1 the colonies are so numerous as to look like fine white dust. In tubes 2 and 3 they become less numerous and larger.

Esmarch's Roll-tubes.—Use liquefied gelatin or agar. The dilutions in tubes 1, 2 and 3 are made as above. Tubes containing a rather small amount of the culture-medium are more convenient. A block of ice should be at hand, and, with a tube filled with hot water and lying horizontally, a hollow of the size of the test-tube should be melted on the upper surface of the ice. In this hollow place the tube of liquefied gelatin or agar; roll it rapidly with the hand, taking care that the culture-medium does not run toward the neck as far as the cotton plug. The medium is spread in a uniform manner around the inside of the tube, where it becomes solidified. Gelatin roll-tubes must be kept in a place so cool that there is no danger of their melting; in handling them they are to be held near the neck, so that the warmth of the hand may not melt the

9

gelatin. Agar roll-tubes should be kept in a position a little inclined from the horizontal, with the neck up, for twenty-four hours, so that the agar may stick to the wall of the tube.

By the plate-method as originally devised by Koch, instead of using Petri dishes, the gelatin was poured upon a sterile plate of glass. This plate of glass was laid on another larger plate of glass, which formed a cover for a dish of ice-water, the whole being provided with a leveling

FIG. 33.

Manner of Making Esmarch Roll-tube.

apparatus. The plate was kept perfectly level until it had solidified, which took place rapidly on the cold surface. The glass plates were placed on little benches enclosed within a sterile chamber. The more convenient Petri dish has displaced the original glass plate to a large extent.

The isolation of bacteria may sometimes be effected by drawing a platinum wire containing material to be examined rapidly over the surface of a Petri dish containing solid gelatin or agar; or over the surface of the slanted culture-medium in a test-tube; or by drawing it over the surface of the medium in one test-tube, then, without sterilizing, over the surface of another, perhaps over several in succession.

Appearance of the Colonies.—The colonies obtained in the Petri dishes or roll-tubes (Fig. 32) may be studied with a hand-lens or with a low power microscope. In the latter case, use the concave mirror with the iris diaphragm partly closed. The colonies present various appearances. Some of them are white, some colored; some are quite transparent and others are opaque; some are round, some are irregular in outline; some have a smooth surface, others appear granular, and others present a radial striation. Surface colonies often present different appearances from those occurring more deeply. Surface colonies are likely to be broad, flat and spreading. If the colony consists of bacteria which have the property of liquefying gelatin, a little funnel-shaped pit or depression forms at the site of the colony. The appearance of colonies may be of great assistance in determining the character of doubtful species. The appearance in gelatin plates of the colonies of the spirillum of Asiatic cholera, for instance, is one of the most characteristic manifestations of this organism.

Pure Cultures.—From these colonies pure cultures may be obtained by what is called " fishing." Select a colony from which cultures are to be made; touch it lightly with the tip of a sterilized platinum wire, taking great care not to touch the medium at any other point. Introduce the wire into a tube of gelatin. Sterilize the wire and plug the tube. In a similar manner, and from the same colony, in-oculate tubes of agar, bouillon, milk, potato and blood-serum. At the same time it is well to make a smear prepa-ration from the colony and to stain with one of the aniline dyes so as to determine the morphology of the bacteria. The growths which take place in the tubes should contain one and the same kind of bacteria. As seen under the microscope their bacteria should have the same general form and appearance as those seen in the colony from which they were derived. This will be the case, provided the colony has resulted from the development of a single bacterium or from several bacteria of the same kind. Occasionally, however, a colony will develop from several bacteria which may not all be alike. In that case a pure culture will not be obtained, and the process of plating may have to be repeated.

CHAPTER VI.

INOCULATION OF ANIMALS.

In the study of pathogenic bacteria, the inoculation of animals is frequently indispensable. The animals most often used are white mice, guinea-pigs, rabbits and pigeons. Larger animals are occasionally employed for special purposes. White mice may be kept in a glass jar covered with wire netting. They may be fed with moistened bread or oats. It is important to see that they receive drinking-water. During inoculation the mouse must be kept in

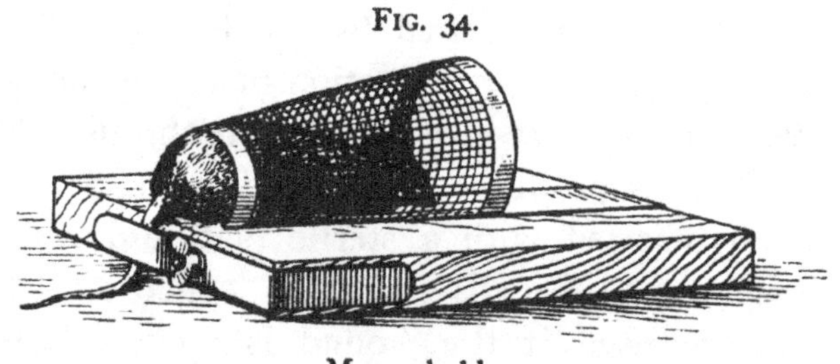

Fig. 34.

Mouse-holder.

position by some sort of mouse-holder, or may be held by an assistant, who takes the skin at the back of the neck between his fingers and at the same time holds the tail. The hair is cut off from the skin at the root of the tail. A small V-shaped opening in the skin is made with scissors, and a stiff sterilized platinum wire is passed into this opening, separating the skin from the muscles for some distance so as to make a pocket. Into this pocket the material is introduced by means of the platinum wire. The

wound may be covered with collodion. The peritoneal
cavity of the mouse may be inoculated with a fluid
culture introduced with a sterile hypodermic
syringe.

FIG. 35.

Guinea-pigs and rabbits, after inoculation, are to
be kept in cages of galvanized iron and wire-
netting. The bottom may conveniently be made in
the form of a movable pan which permits of the
disinfection of the excreta. Rabbits and guinea-
pigs may be fed with oats, carrots, cabbage, grass
and the like. Guinea-pigs and rabbits may be held
by an assistant or tied by the legs upon a board.
The hair over a small portion of the abdomen is cut
away and a short incision is made through the skin:
a pocket is produced with a stiff wire, and the material inserted with a sterile platinum wire. The
wound may be covered with collodion. Sutures
may be used if the wound is large. Solid substances may conveniently be introduced by placing
them in a sterile glass cannula, which is pushed to
the proper situation through a small incision. The
substance in the cannula is forced out of it with a
stiff sterile platinum wire. (Fig. 35.) The peritoneal cavity may be inoculated with a previously
sterilized hypodermic syringe, or an incision may
be made which reaches to the peritoneal cavity,
into which the desired substance may be introduced
with a sterile platinum wire, the incision being
closed with sutures.

Intravenous inoculation is most commonly practiced upon rabbits. A small vein which is near the
posterior margin of the ear of the rabbit is easily
reached from the dorsal surface; the hypodermic
needle is introduced directly into this vein. In making a

hypodermic injection, the needle and syringe should of course be sterilized before and after each operation.

Autopsies upon animals should be held as soon as possible after death. During the interval the body should be kept in the ice-box. The autopsy room should be furnished with screens to keep out flies, so that they may not light on the infected animal. The animal should be extended on its back upon a board. The legs may be fastened with pins or tacks. The animal should be handled with forceps as far as possible, and after beginning the autopsy the fingers should not touch it. If the fingers come in contact with infectious matter, disinfect them at once. Have a basin of bichloride of mercury solution 1–1000 ready for this purpose. Knives, scissors, platinum wires and forceps should be sterilized in the flame before and after each manipulation. Be prepared to make smear preparations on cover-glasses, and to inoculate tubes of gelatin, agar and other media as desired. Moisten the hairs over the thorax and abdomen with bichloride of mercury solution 1–1000, to prevent them from being carried into the air. Make an incision, passing through the skin from the sternum to the pubis along the thorax and abdomen, and diagonal incisions extending down the fore and hind legs. Dissect away the skin from the thorax, abdomen, and upper parts of the legs. With a knife heated in the flame, sear a broad line extending down the middle of the abdomen. Through this burned surface make an incision through the muscles of the abdomen. In a similar manner make a transverse incision across the middle of the abdomen through a burned surface. Cultures should be made from the peritoneal cavity, and smears upon cover-glasses prepared, which are afterwards to be stained. With a hot knife, scorch a small area on the surface of the liver; through this surface enter the liver with a sterilized platinum wire, and with the material withdrawn inoculate the tubes; also make cover-glass preparations. In the same manner inoculate tubes and make cover-glass preparations from the spleen, the kidneys, the pleural cavity, the pericardial cavity, the lungs, and the blood inside the heart. All incisions are to be made through the burned surfaces, and all material collected for inoculation is to be obtained through burned surfaces. In sterilizing the instruments in the flame avoid sputtering, especially when they become covered with oil from adipose tissue. Pieces of lung, liver, spleen, kidney and other organs, as may be indicated, should be placed in 95 per cent. alcohol for fixation and hardening. The animal and the board on which it was extended should be covered with bichloride of mercury solution 1–1000, and afterwards burned. The cage or jar and the instruments, dishes and towels used should be sterilized by steam. The hands of the operator should be washed thoroughly with soap and water and with a 1–1000 solution of bichloride of mercury.

FIG. 36.

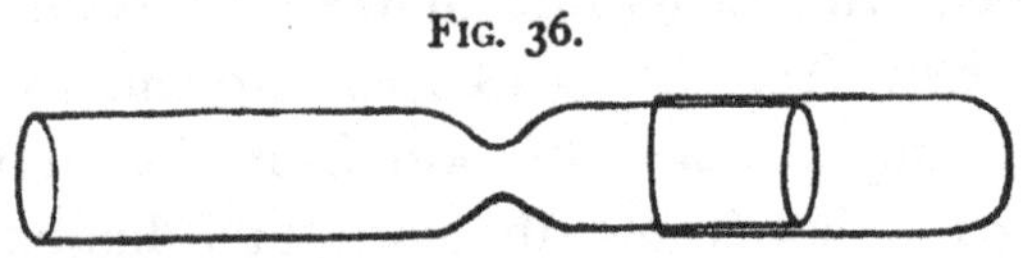

Method of Making Collodion Capsules. (After McCrae.)

Collodion Capsules.—Bacteria may be cultivated in the living body of an animal, without infecting the animal, when they are enclosed in collodion capsules. Their soluble products are able to diffuse through the collodion, while the animal's fluid may pass into the sac to nourish them. These capsules were originally made by dipping the round end of a glass rod into collodion repeatedly. McCrae's method[1] is easier and more satisfactory. (Fig. 36.)

A piece of glass tubing is taken, and a narrow neck drawn on it near one end. This end of the tube is rounded in the flame, and the body

[1] *Journal of Experimental Medicine*, Vol. V., p. 635.

of a gelatin capsule is fitted over it, while still warm, so that the gelatin may adhere to the glass. The capsule is now dipped into 3 per cent. collodion, covering the gelatin and part of the glass. It is allowed to dry a few minutes, and is dipped again. In all two or three coatings may be given. The capsule is filled with water and boiled in a test-tube with water. The melted gelatin is removed with a fine pipette. The capsule is partly filled with water or broth and sterilized. The capsule may now be inoculated. The narrow part of the neck must then be sealed in the flame, taking care that the neck be dry. The sealed capsule should be placed in bouillon for twenty-four hours. No growth should occur outside the capsule if it is tight. It may now be placed in the peritoneum of an animal.

CHAPTER VII.

COLLECTION OF MATERIAL.

SAMPLES of water or milk collected in sterilized tubes or bottles, when they are not examined immediately, or when they are to be transmitted any distance, should be kept on ice. Specimens of sputum may be collected in clean bottles tightly corked. They should be examined as soon as possible. Although decomposition appears not to interfere with the staining properties of the tubercle bacilli, the sputum should be fresh in order that the other bacteria contained in it may be studied. Therefore it should be free from contamination with putrefactive germs. Valuable information can also be obtained by examination of sputum in a fresh condition before staining (see also page 44).

Samples of urine keep better after the addition of a few crystals of thymol, which retards the fermentative process, so that the sedimentation of the bacteria and of other solid matter in conical vessels is facilitated, although that purpose can be accomplished at once by the centrifuge. Thymol will also be a useful addition, as far as a bacteriological examination is concerned, in case samples of urine are to be sent by mail; thymol should not be added if cultures are to be made.

Specimens of sputum, pus or blood may be collected conveniently in the form of thin smears upon cover-glasses. The smears are fixed by passing through the flame three times. Smears of blood are prepared as follows: Have two perfectly clean, square cover-glasses. The finger, or the lobe of the ear, having been carefully washed with water, alcohol, and ether, is punctured with a sterilized needle, and a small drop of blood issues which is wiped away. The second drop of blood should be taken; it should be about the size of a pin's head. No pressure should be exerted upon the skin. This drop of blood is placed on one of the cover-glasses. The other cover-glass is laid upon the first, both being handled with forceps. The drop of blood becomes flattened out into a thin film. Immediately and before the blood has had time to coagulate the two are slipped or slid away from each other in a horizontal plane, not forcibly pulled apart. The blood, therefore, will be spread in thin films on the cover-glasses. It is best to place the cover-glasses so that one does not cover the other exactly, but so that the sides of the one lie diagonally to the sides of the other, although their centers coincide (Fig. 37). Films of blood which are to be examined for the parasite of malaria may be prepared in this manner. Samples of blood to be used for the serum reaction for typhoid fever need to be pretty good-sized drops of blood, which may be collected on cover-glasses or pieces of un-sized paper and allowed to dry. To test blood by culture methods, 1 to 5 c.c. may be drawn from a vein during life, using a sterilized hypodermic syringe and all antiseptic precautions. The blood thus taken may then be used for cultures in various ways. A good method for general purposes is to empty the syringe quickly into a flask holding 100 c.c. or more of bouillon or dextrose-bouillon. The mixture of blood and bouillon should be placed in the incubator for one to two days. If the bacteria develop, they may be secured in pure cultures by plating, and may be studied further, as the occasion requires.

FIG. 37.

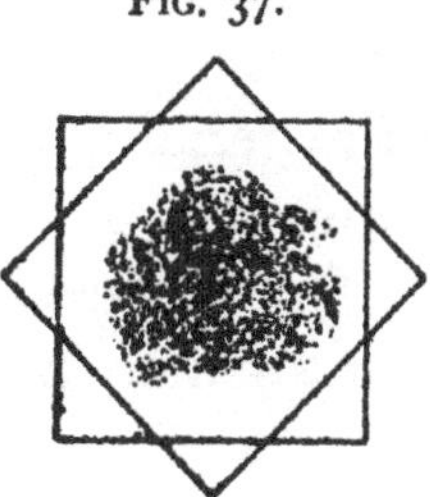

Manner of Placing Cover-glasses in Making Films of Blood. (After Cabot.)

At autopsies on human subjects plate-cultures should be made, if possible, directly from the organs. In all cases organs should be entered by the platinum wire through burned surfaces. The method of isolation by streaking the platinum wire containing the material under examination lightly, several times, over the surface of an agar plate, will be found convenient. At the same time smears should be made from the organs upon cover-glasses for microscopical study, and portions of the organs should be saved and hardened in alcohol.

A convenient device for the collection of infected material is a stiff wire wound with a pledget of absorbent cotton at one end, the whole sterilized in a tube, as recommended by Warren for collecting pus and other fluids for examination, and as introduced by W. H. Park for the collection of material from the throat in cases of suspected diphtheria (Fig. 78).

The so-called Sternberg bulb is valuable for the collection of fluid materials for examination. A short piece of glass tubing is taken; at one end is blown a bulb; the other end is drawn out to a long, fine point. To introduce the substance into the bulb, the expanded end is heated in the flame; the point is broken and introduced below the surface of the fluid which is to be collected; as the bulb cools, the air in it contracts and draws the fluid into it. When it has taken up as much as it will, the point may again be closed in the flame.

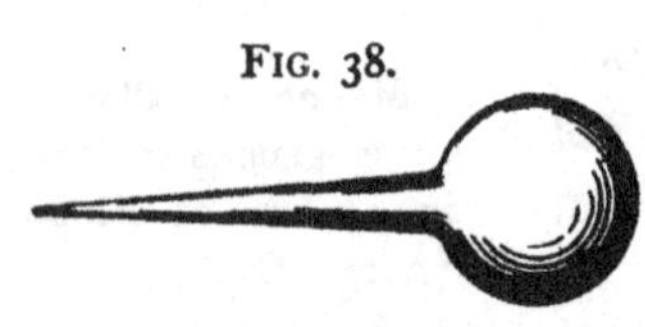

FIG. 38.

Sternberg Bulb.

When infectious material is to be transported, it should be so packed that breakage or leakage is impossible.

Concerning the transmission of materials containing bacteria in the mails, the ruling of the post-office department of the United States, March 2, 1900, is as follows:

"That the order of the Postmaster General of December 27, 1897 (Order No. 677), amending Order No. 88 of February 5, 1896, prescribing the conditions under which specimens of diseased tissues may be admitted to the mails is hereby further modified in the following manner:

"Specimens of diseased tissues may be admitted to the mail for transmission to United States, State, or municipal laboratories, only when enclosed in mailing packages constructed in accordance with the specifications hereinafter enumerated: Liquid cultures, or cultures of microorganisms in media that are fluid at the ordinary temperature (below 45° C. or 113° F.) are unmailable. Such specimens may be sent in media that remain solid at ordinary temperatures.

"Upon the outside of every package shall be written or printed the words 'Specimen for Bacteriological Examination. This package to be treated as letter mail.' No package containing diseased tissue shall be delivered to any representative of any of said laboratories until a permit shall have first been issued by the Postmaster General certifying that said institution has been found to be entitled, in accordance with the requirements of this regulation, to receive such specimens."

The regulation includes not only cultures but "moist specimens of diseased tissues." The specifications prescribing the manner of packing, which are minute and complicated, may be obtained from local postmasters.

CHAPTER VIII.

SYSTEMATIC STUDY OF SPECIES OF BACTERIA.[1]

In order to conduct the study of any species of bacteria it is necessary to have the organism isolated in a pure culture by the plate-method, or by some other method already described. Having thus obtained the organism in pure culture, it is to be examined with reference to its behavior in certain particulars. It is well for the beginner to study a few known species of saprophytes obtained from some reliable laboratory in pure culture. The points which are to be considered can be illustrated best by presenting them in tabular form, filling out the items of the table for a given species of bacteria.

1. Name.
2. Habitat or source.
3. Morphology; grouping, as in chains or in zoöglœæ.
4. Size.
5. Staining proporties. Behavior by Gram's Method.
6. Capsule, present or otherwise.
7. Spore formation.
8. Motility, flagella.
    Growth on culture-media.
9. Relation of growth to temperature.
10. Gelatin; observe whether the gelatin is liquefied or not. Colonies in gelatin plates, study under low power of microscope.

[1] For the identification of unknown species consult "A Manual of Determinative Bacteriology," by Frederick D. Chester.

11. Agar.   Colonies in agar plates, study under low power of microscope.
12. Bouillon, note cloudiness, pellicle, or precipitate.
13. Milk; observe whether or not the milk is coagulated and subsequently peptonized.
14. Production of gas in fermentation-tube with bouillon containing sugar, as dextrose, or in agar with sugars.
15. Potato.
16. Blood-serum; observe whether or not peptonization occurs.
17. Production of indol.
18. Pigment formation.
19. Production of acid or alkali.
20. Relation to oxygen; observe whether the superficial or the deep part of the growth is the more luxuriant in stab-cultures; use anaërobic methods if necessary.
21. Pathogenesis.

In commencing the study of bacteriology the pupil should try the common staining methods and make the most important culture-media.   Having culture-media prepared, it is customary to study a number of species of non-pathogenic bacteria.   Notes of the work and sketches showing the morphology of the organisms should be made.   It is well to choose species which have properties decidedly different from one another.   The micrococci, bacilli and spirilla should be represented; forms that are motile and that are not; species that form spores and others that do not form spores; some that liquefy gelatin and some that do not. There should be chromogenic forms, and species that ferment dextrose, and that produce indol,—such species as some of the sarcinæ, the bacillus coli communis, the hay bacillus, the potato bacillus, bacillus prodigiosus, a bacillus fluorescens and spirillum rubrum. It is well, when possible, to obtain material directly from nature rather than from laboratory cultures. This may readily be done in the case of the hay bacillus and the potato bacillus. Fecal matter may be spread on gelatin plates and the bacillus coli communis obtained in pure culture. Fluorescing bacilli are very common in water. Large spirilla are often found in swamp water. Some organisms like spirillum rubrum can only be had from laboratory cultures. The growth of some aërobic organism, like the potato bacillus, may be tested under a cover-glass (see Fig. 29). The pyogenic bacteria, which can easily be isolated from pus, may be studied in this connection with great advantage. The staphylococcus pyogenes aureus and the streptococcus pyogenes should on no account be omitted. The diplococcus of pneumonia can most readily be obtained from a mouse or a rabbit which has died with pneumococcus infection. Such an animal can best be infected by subcutaneous inoculation, using some of the rusty sputum of a case of lobar pneumonia. The cultivation of the pneumococcus will be found to present difficulties in classes containing large numbers of students.

Representative forms of moulds and yeasts should be studied at the same time. Moulds are easily obtained by exposing some nutrient substance to the air, covering it, and allowing cultures to develop; yeasts will probably grow also. Ordinary brewer's yeast may be isolated in pure culture from gelatin plates. Bacteriological examinations also should be made of air, soil, water and milk. With such simple means, all the important properties of bacteria may be demonstrated.

Experiments in sterilization and disinfection as described in Chapter VIII., Part II., may be performed with the bacteria mentioned, which present every variety of resisting power up to the almost incredible toughness of the spores of the hay and potato bacilli. The efficiency of the methods used for sterilizing surgical materials, as silk and catgut (Chapter IX., Part II.), should be tested; also, of the methods for disinfecting the hands; if possible, of the methods for disinfecting rooms, as well.

After some proficiency has been acquired, various pathogenic bacteria may be studied as the circumstances of the case require. Much judgment has to be used in allowing students to work with pathogenic bacteria. Anthrax, glanders, tetanus, cholera, bubonic plague, Malta fever, and diphtheria all have occurred in laboratory workers through accidental infection, sometimes with fatal results. The various rules for the management of the platinum-wire, hanging-drop slides and sputum bottles, and for the handling of cultures and other infectious materials have already been given (pages 33, 36, 44 and 84 to 88). The most important precaution, perhaps, is observance of the rule that while working in the laboratory, nothing should be put in the mouth. Cultures should never be carelessly left in improper places. Cultures of bacteria should be thoroughly sterilized before the tubes are cleaned. The writer is in the habit of having tubes and dishes containing pathogenic bacteria placed in the steam sterilizer for an hour on each of three days, and of having the plugs removed and burned and the tubes filled with 5 per cent. carbolic acid between the second and third sterilizations. In taking these measures, the same kind of reasoning applies as that which induces engineers to give bridges from four to six times the strength they need to bear the greatest strain likely to be put upon them, or to make the boiler of a steam engine strong enough to bear six times the greatest pressure which it is expected that the steam contained in it will exert.

PART II.

CHAPTER I.

CLASSIFICATION; GENERAL MORPHOLOGY AND PHYSIOLOGY OF BACTERIA.

THE relationships existing between bacteria and other kinds of organisms are not perfectly clear. It is quite generally conceded, however, that bacteria are plants. They show affinities with both the lower algæ and the lower fungi, and they have even some points of resemblance with certain of the protozoa. On account of their extreme smallness it is impossible to analyze the structure of the individual bacteria and to contrast the structure of one with that of another. The classification cannot therefore be established on morphological grounds chiefly, as is done with large animals or plants. We are obliged to rely also upon their growth with relation to the presence or absence of oxygen and to temperature, their behavior on culture-media, the appearances of the growths, and the production of certain substances with peculiar chemical reactions, when we wish to establish the points of difference between one species and another—all of which is extremely unsatisfactory and probably not perfectly reliable. It is likely that forms which are now considered as different species are not really such in all cases, and also that different species may now be included under one heading as a single species. Notwithstanding the unsatisfactory condition of the classification of bacteria, it must not be supposed that the species of bacteria are not permanent. For instance, it would be incorrect to imagine that it is possible for the micrococci and spirilla to become converted into species of bacilli, or for the bacilli of one species to be transmuted into those of another. This does not contradict the statement that we may frequently, through erroneous and imperfect information, be in the habit of including unlike species under one name, or of classifying mere varieties of one species as entirely different species. At present the simple division of bacteria into three great generic groups is probably as good as any: *micrococci,* spherical forms; *bacilli,* rod-shaped forms, one diameter being in excess of the others; *spirilla,* twisted like a corkscrew, making long spirals or simply parts of spirals (comma-shaped forms).

Recent investigations indicate that several species of bacteria often are closely related to one another, so as to form

FIG. 39.

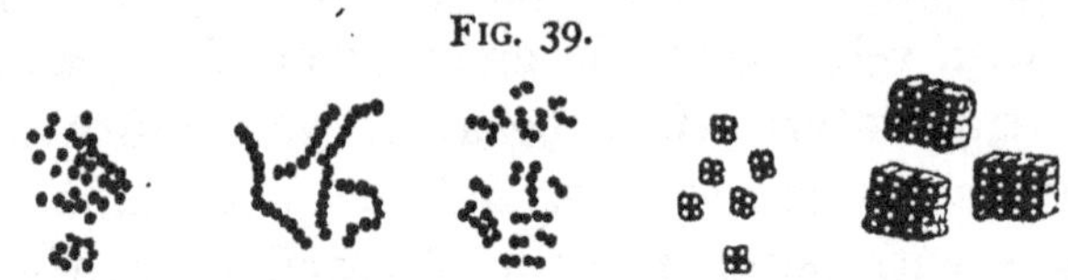

Staphylococci. Streptococci. Diplococci. Tetrads. Sarcinæ.

a well-marked group. Such a group is constituted by the bacillus of typhoid fever, bacillus coli communis and similar forms. The spirillum of cholera and other comma-shaped spirilla resembling it may be held to constitute another group. Still another is that containing the tubercle bacillus and other acid-proof bacilli.

The micrococci are subdivided into *staphylococci,* where the spheres grow in clusters like a bunch of grapes; *streptococci,* where they are arranged in long rows or chains, like a string of beads; *diplococci,* or pairs of micrococci; *tetrads,* where the individual spheres are grouped in fours; *sarcinæ,* where they are grouped in eights, making the outline of a cube, resembling a bale or package tied with rope.

The bacilli are not usually subdivided in this manner, although their forms vary considerably. The ends are sometimes square, sometimes round. Sometimes they are very short. Sometimes they grow in longer, thread-like forms, in which, however, the transverse markings which

FIG. 40.

Bacilli of Various Forms.

indicate the outlines of the individual bacilli can generally be seen, and which resemble a bamboo rod. Short oval bacilli may look exceedingly like micrococci. Bacilli with rounded extremities, placed end to end, look like strings of sausages. Under exceptional circumstances, branching forms of the bacilli of diphtheria, tuberculosis, glanders and bubonic plague and various other species have been encountered.[1]

The word "*bacterium*" was formerly used to designate short bacilli which generally formed no spores, while the

FIG. 41.

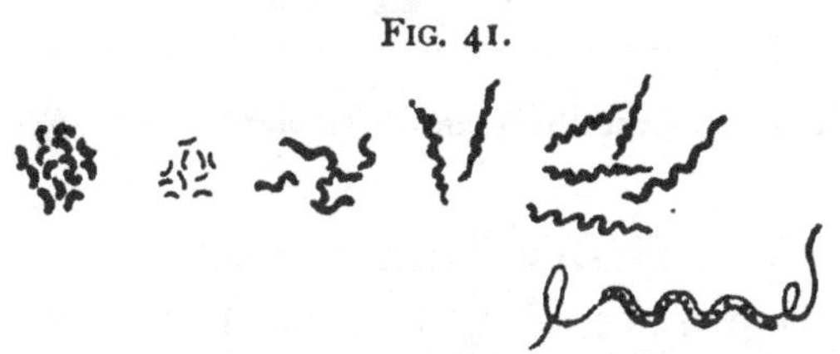

Spirilla of Various Forms.

word bacilli was restricted to the longer forms in which spore formation occurred. This use is no longer common, although not rarely the name bacterium is still given to a species—for instance, bacterium coli commune.

[1] See Hill, *Journal of Medical Research*, Vol. VII., January, 1902; Loeb, *ibid.*, Vol. VIII., 1902.

Spirilla present a very great variety of form. The short "*comma-shaped bacilli*" are only parts, at most, of spirals, although the microbes of cholera do sometimes form long spirals. On the other hand, there are among spirilla large and long sinuous figures which present most remarkable pictures under the microscope; for example, the spirillum of relapsing fever. Formerly spirilla without very marked windings were called "*vibrios*"; and long, wavy forms with corkscrew-like windings "*spirochætæ*"; and the rigidly spiral forms were denominated "*spirilla.*" These definitions have for the most part lost their significance, although the names still linger in nomenclature.

FIG. 42.

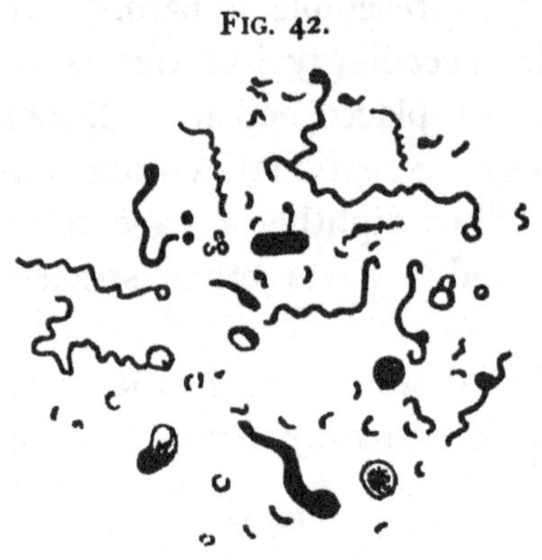

Involution Forms of the Spirillum of Cholera. (Van Ermengem.)

Besides the classification already mentioned, bacteria are sometimes grouped according to certain other qualities. In general botany, saprophytes are plants that grow on decaying vegetable matter. In a bacteriological sense, *saprophytes* are bacteria which grow in external nature on dead organic matter, and *parasites* are bacteria which exist upon the living tissues or fluids of any organism. Nearly synonymous with the above words are those which do not and those which do produce disease, or *non-pathogenic* and *pathogenic.* The adjectives *facultative,* or optional, and *obligate,* or strict, are used to qualify the above terms and many others.

Size.—Bacteria vary greatly in size. The micrococci are usually $1\,\mu$ or less in diameter. The short diameters of bacilli and spirilla also are less than $1\,\mu$ as a rule, while the length may be several μ . The anthrax bacillus ($1.5\,\mu$ $\times$ 3 to 10 μ) and the spirillum of relapsing fever are the largest bacteria known to be pathogenic to man. To say that a micrococcus is $1\,\mu$ in diameter means that 25,000 end to end would make a line 1 inch long. It has been estimated that 1 milligram of a pure culture of the staphylococcus pyogenes aureus contains 8,000,000,000 micrococci.

There is good reason for believing that organisms exist, which are too small to be visible with the most powerful microscopes. The nature of these organisms is not known, but it is not improbable that some of them are bacteria. (See pleuro-pneumonia of cattle, etc., Part II., Chapter V.)

In stained preparations the bodies of bacteria frequently seem to be homogeneous. On the other hand, they may exhibit certain spots which stain more intensely than others, the stained spots alternating with clear areas. The dark-staining granules may take a slightly different shade of color from the rest (metachromatic granules, Babes-Ernst bodies). Somewhat similar appearances may result from changes in the density of the protoplasm of bacteria, leaving vacuoles that do not stain (plasmolysis).

In old cultures bacteria are likely to show deformed and twisted outlines called *involution forms.* It is not uncommon for bacteria to be enclosed in a kind of envelope of some clear substance, which stains with difficulty or not at all, called a *capsule.* The paired micrococci of pneumonia are inclosed in such capsules. The capsule is more likely to be demonstrated when the bacteria are obtained from the fluids derived from an animal's body than when they have been grown artificially in culture-media. A *zoöglœa* is a large mass of bacteria in a resting condition held together by a mucilaginous substance. The composition of bacteria varies considerably with different species. The basis appears to be proteid substance.

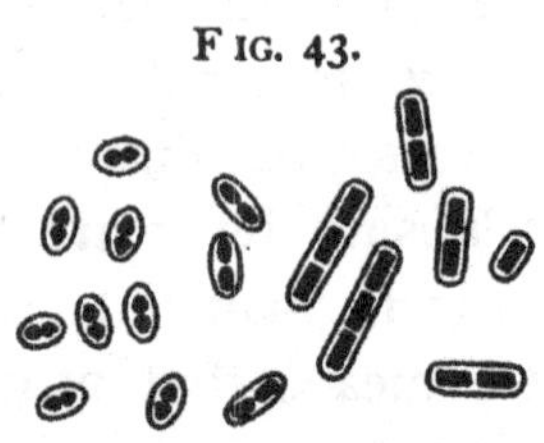

F IG. 43.

Bacteria with Capsules.

The multiplication of bacteria takes place in almost all cases by transverse fission. The formation of tetrads or sarcinæ from micrococci depends upon fission in two or three planes. Repeated fissions of micrococci in one plane result in the formation of streptococci. Micrococci that have recently divided are likely to be somewhat flattened. Multiplication under favorable circumstances may take place at a phenomenally rapid rate. Bacilli have been observed to divide in twenty minutes. If division takes place once in an hour, the progeny of one organism at the end of twenty-four hours will be 16,777,216, *i. e.*, $(2 \times 1)^{24}$. The ordinary form of reproduction by fission is called *vegetative,* and bacteria that are multiplying in this manner are often spoken of as being in the vegetative condition.

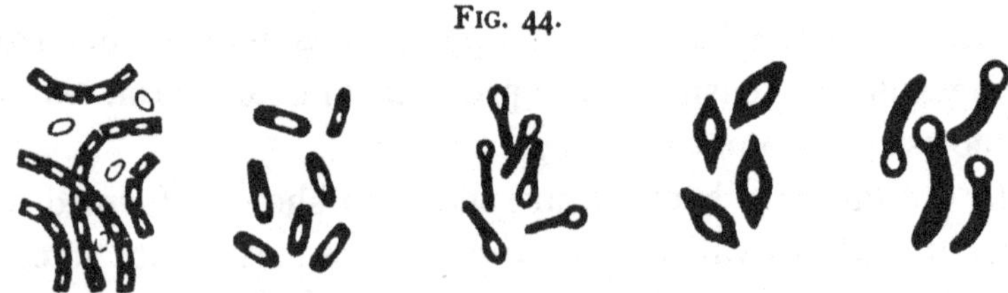

FIG. 44.

Bacteria with Spores.

Spores.—Under certain circumstances the reproduction of bacteria takes place by means of bodies called *spores.*

They appear in a typical form in the large bacilli, where, near the centers of the bacilli, highly refracting, shining spots may be seen which are found to stain less rapidly with the aniline dyes than the rest of the bacilli.   They are not to be confused with the unstained spots described as vacuoles.   On account of their being formed from a part of the interior of the bacterium, such spores are called *endogenous*.   These spores are found mostly in the bacilli, rarely in spirilla and micrococci.   They are what is meant when the word spore is used alone without qualification. The existence of another kind of spore, described as forming from the whole of the bacterium (called *arthrospore*), is doubtful.   At all events, its significance is not at present understood.   Spores develop generally, though not always, under adverse conditions of various kinds, as of temperature and of nutrition.   They are more resistant to unfavorable influences of all sorts than are the fully developed bacteria.   Spores resist drying, light, heat and chemical agents to a remarkable degree, at times.

Anthrax spores are said to have been found which could withstand steam for twelve minutes, 1–1000 mercuric chloride for nearly three days, or 5 per cent. carbolic acid for more than forty days.   The greatest resistance is displayed by the spores of some of the saprophytic bacteria, particularly those of hay and potato, which are sometimes not destroyed by several hours of steaming; and bacteria are said to have been obtained from the soil which resisted 100° C. for sixteen hours.   When cultivated at a temperature as high as 42° C. the anthrax bacillus becomes incapable of forming spores.   Spores themselves do not multiply, nor do they manifest any activity.   Spores may be located at the center of the bacillus, or nearly at one end, when the end of the bacillus is likely to enlarge, making a form having the shape of a drumstick, as takes place with tetanus bacilli (Fig. 44). When a bacillus assumes a spindle shape on account of having the middle part bulged through the formation of a spore it is called a *clostridium*. With rare exceptions, a single bacillus contains but one spore. Under favorable conditions the spores germinate, as it is called, and develop to the adult form of the organism. This may be witnessed in hanging-drop preparations.

Motility.—Motility is rarely exhibited by micrococci; some bacilli possess it and some do not; while nearly all of the spirilla are motile. The phenomenon is observed in the hanging-drop. The degree of motility is variable, being

FIG. 45.

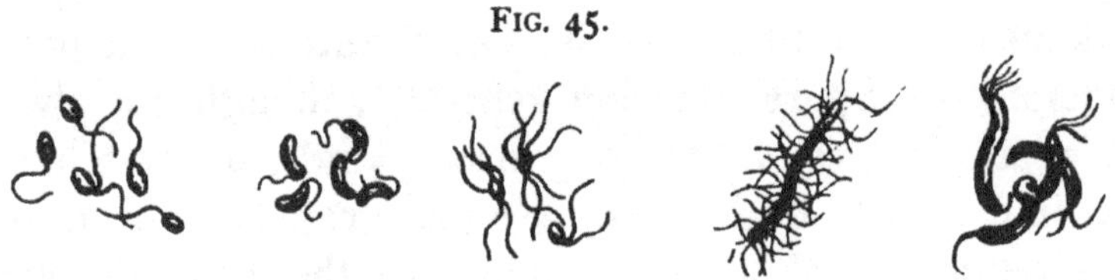

Bacteria showing Flagella.

sometimes slight and sometimes very active. When seen under a high power the little particles taken from a culture of a motile organism may look like a writhing mass of maggots or like tadpoles in a pool. The motility is most active in young cultures. The movement results from the vibration of little processes, or *flagella* (Fig. 45). Of these there may be one or several. They may be placed singly or in groups, at the end, or scattered around the sides. They are extremely difficult to demonstrate except by special staining methods, which, furthermore, are decidedly capricious. After the flagella have been stained, the bacteria appear somewhat larger than when stained by the ordinary methods. The flagella upon the bacilli of typhoid fever are numerous and form a very striking picture.

Chemotaxis.—Motile bacteria possess the property of being attracted by certain substances (positive chemotaxis) and of being repelled by others (negative chemotaxis). Similar properties are widely distributed among living cells, both animal and vegetable.

CONDITIONS FAVORABLE FOR THE GROWTH OF BACTERIA.

Warmth.—Among the different kinds of bacteria forms are said to exist which multiply at temperatures as low as 0° C., while there are species that multiply at 70° C. Bacteria which flourish at a very high temperature (maximum about 70° C.) are called *thermophilic.* The pathogenic bacteria usually flourish better at a point somewhere near the temperature of the human body. This is not necessarily the case with the non-pathogenic species. Ordinary water bacteria thrive better at ordinary temperatures.

Sternberg's method for determining the *thermal death-point* of a species of bacteria is to draw portions of a pure culture of the organism into capillary tubes with expanded ends, when the tubes are sealed in the flame. The tubes are supported upon a glass plate placed in a water-bath, whose temperature is indicated by a thermometer, while a uniform temperature is secured by stirring. The time of exposure is, as a rule, ten minutes. The tubes should be removed quickly to cold water. Their contents should afterwards be inoculated into bouillon to determine whether or not the organisms have been killed.

Moisture is indispensable to the growth of bacteria, and drying causes the death of certain kinds, as, for instance, the spirillum of cholera.

Food.—There are a few species of bacteria that contain chlorophyll, but it is wanting in most forms. On account of the absence of chlorophyll, bacteria require, as part of their food, organic compounds containing carbon, such as sugar. They are unable, with possibly a very few exceptions, like the nitrifying bacteria, to derive their carbon from the carbon dioxide of the atmosphere, or from inorganic carbon compounds. Although some species are able to obtain nitrogen from inorganic salts, most bacteria flourish best if organic substances containing nitrogen, like peptone and albumen, are furnished them as part of their food. The complicated, unstable, organic molecules with high potential energy are converted by them into simple and more stable compounds like carbon dioxide, ammonia and water, with the liberation of energy. These facts become manifest in connection with their important work in decomposition, putrefaction and fermentation. A culture-medium having a slightly alkaline or neutral reaction is favorable to most bacteria.

The prolonged artificial cultivation of bacteria may or may not modify their properties. The pathogenic bacteria are likely to undergo considerable modification both in the quality and luxuriance of their growth and the intensity of their pathogenic characters.

The growth of bacteria may eventually be hindered by the accumulation of the products of their own metabolism. Many bacteria refuse to grow on culture-media at all. Some species are extremely fastidious, and can only be propagated on particular sorts of nutrient substances.

Relation to Oxygen.—Oxygen is indispensable to the growth of some bacteria, *aërobes*. Its absence is equally indispensable to certain others, *anaërobes*. Others still are able to flourish either in the presence or absence of oxygen, facultative aërobes or anaërobes. The first-named varieties are sometimes called strict, or obligate aërobes or anaërobes.

Effects of Sunlight.—Direct sunlight kills the vegetative forms of bacteria more or less rapidly, and constitutes one of the most efficient among the natural methods of disinfection. Diffuse daylight acts much more slowly. Electric light acts like sunlight or daylight, the results being dependent on the intensity of the light. The violet part of the spectrum is most active.

The influence of *electricity* upon bacteria has not yet been fully studied. Apparently the destruction of bacteria reported as having been effected by electricity was the result of electrolysis of the medium.

It appears probable that *X-rays* do not produce important effects on bacteria, although further investigation of this subject is needed. The success which has attended the use of light rays and X-rays in the treatment of lupus and other diseases is not necessarily to be explained as the result of bactericidal action of the rays.

CHAPTER II.

PRODUCTS OF THE GROWTH OF BACTERIA.

Phosphorescence.—Bacteria whose cultures exhibit phosphorescence have been found in the ocean and in fish.

Chromogenic Bacteria.—Many bacterial growths display brilliant coloring. The different species of sarcinæ are remarkable for forming highly-colored growths; some of them are rose-red, some orange-yellow, some lemon-yellow, and so on. The bacillus prodigiosus presents a brilliant red growth whose rapid development is said to have formed the basis for the so-called " Miracle of the Bleeding Host " (see page 15). The bacillus pyocyaneus in culture gives a brilliant green fluorescence and is responsible for the color of blue or green pus.

Bacilli which exhibit a green fluorescence in cultures are common in water. In cultures on potato or agar the colors of the chromogenic forms are usually well shown.

Ferments or Enzymes.[1]—Many bacteria form ferments which have the power of dissolving proteid substances in a manner similar to trypsin. The liquefaction of gelatin is a familiar example of this process. The property of liquefying gelatin, or otherwise, is used in classifying bacteria and in determining the nature of unknown species.

Some bacteria, as the bacillus coli communis, form ferments which act like rennet in coagulating milk. Other bacteria are capable of forming sugar from starch. Others have the power of changing cane-sugar into glucose.

[1] Consult Buxton, " Mycotic Enzymes," *American Medicine,* July 25, 1903.

Bacteria which are able to decompose cellulose are found in the stomachs of ruminant animals. Although it is doubtful whether the products of cellulose decomposition have any nutritive value, the process is useful in effecting a subdivision of the coarse food, consisting of grass, hay, and the like.

Some bacteria have the power of decomposing neutral fats into fatty acids and glycerin, after the manner of the fat-splitting ferment of the pancreatic juice.

The *end-products* which result from the growth of bacteria upon albuminous nutrient media are very numerous. They are complicated and not well understood. Among these end-products may be mentioned peptone, indol, skatol, phenol, leucin and tyrosin. Nearly related are the toxins, which play an important part in the production of disease by pathogenic bacteria. In the decomposition of urine by bacteria the urea is converted into ammonium carbonate.

The *formation of indol* in cultures is an important peculiarity of certain bacteria, which may be tested as follows: The bacteria are cultivated in Dunham's peptone solution or in dextrose-free bouillon; after twenty-four to forty-eight hours the test may be made. Add ten drops of concentrated sulphuric acid; the development of a rose-color indicates the presence of both indol and nitrites. If no rose-color forms, to another tube add, first 1 c.c. of a 0.01 per cent. solution of sodium nitrite, and then the sulphuric acid. The development of a rose-color indicates the formation of indol but not of nitrites. If there is no rose-color, no indol has been formed. The color appears usually in a few minutes, but it may only develop after a somewhat longer time. Control tests must be made upon tubes of the same peptone solution but which have not been inoculated. The reaction may be hastened by warming slightly. The value of this reaction will be understood when, to give one illustration,

it is remembered that the bacillus coli communis produces
indol and the bacillus of typhoid fever usually does not.
The reaction depends upon the liberation of nitrous acid,
which, with indol, forms a red color.

The change of organic substances into more stable ones
does not cease with the compounds mentioned above. Certain bacteria of the soil which will be discussed further on
are able to complete the conversion of ammonia into nitrous
acid (leading to the formation of nitrites) ; and others still
that of nitrites into nitric acid, which at once forms nitrates.

Formation of Acids.—In the course of their growth
many bacteria produce acids, especially from substances
containing sugar. The power of developing lactic acid is
possessed by a large number of species. Acetic acid is
another common by-product. Besides these, butyric acid,
formic acid, propionic acid and many more are formed by
different bacteria.

Development of Gas.—The evolution of gas from bacterial growths is of frequent occurrence. Carbon dioxide,
hydrogen sulphide and nitrogen are among the better
known gases that may be formed. The odors that arise
from cultures and that are so characteristic of putrefactive
processes depend upon the development of gases, or a mixture of gases, of considerable complexity. The bacillus
aërogenes capsulatus leads sometimes to the formation of
gas in the organs of the human cadaver within a short time
after death.

T. Smith considers the formation of gases in media containing sugar of importance in discriminating between different species. Bouillon containing 1 per cent. of dextrose
(or lactose, etc.) is the culture-medium advised. The
test is best conducted in a U-shaped tube, closed at one end,
and at the other end provided with a bulb (Fig. 46). The
tube is stoppered with cotton, sterilized by dry heat, after-

ward filled with the bouillon, and sterilized by steam in the usual manner. After the last sterilization it should be tilted until the closed end is completely filled with the medium. After it has been inoculated with the species under consideration, any development of gas will be indicated by the collection of the gas at the closed end. The amount of gas formed may be estimated and its quality tested. To accomplish the latter fill the bulb with 2 per cent. solution of sodium hydroxide, close the outlet, and tilt the tube to allow the mixture to come in contact with the gas. After shaking, this causes the absorption of the carbon dioxide and diminution in the quantity of gas. The portions which remain may be mixed with air and ignited, when the presence of hydrogen and some of its compounds will be indicated by an explosion. (See The Detection of Bacillus coli communis in Water, Part II., Chapter III.)

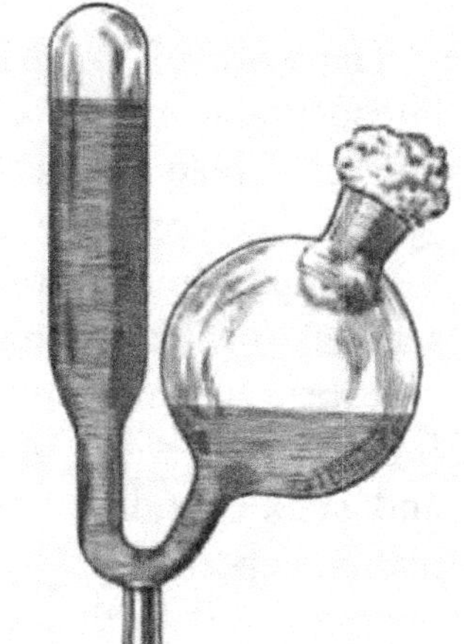

FIG. 46.

Fermentation-tube.

The development of gas may readily be tested by inoculating the bacteria by a deep puncture into agar containing 1 per cent. of dextrose or other sugars. The development of gas causes bubbles to form in the agar, often to the extent of splitting it, and sometimes forcing out the cotton plug (see Fig. 68).

The activities of bacteria which have just been enumerated are fundamental to the phenomena which go by the names of *fermentation* and *putrefaction*. These words have been defined differently at different times and by different writers, but in general both are used as names for the breaking up of complex organic compounds by micro-organisms with the formation of simpler compounds. Fermentation refers especially to the formation of useful products like alcohol. The term putrefaction is employed chiefly for the breaking up of nitrogenous compounds with the development of foul-smelling gases. The term fermentation is also applied to the decomposition of complex substances through the influence of unorganized ferments or enzymes.

The work of bacteria in fermentation and putrefaction is indispensable to the existence of the organic world as we find it. Green plants convert the stable compounds of nitrogen, the carbon dioxide of the atmosphere, and water into the complex and unstable albumins and carbohydrates which serve as food for animals. Animals, on the other hand, convert these unstable and complex compounds back into simpler forms. The work of changing them back into the simple and stable condition, in which they serve as the food for plants, is performed by animal life in part only, and its completion is left to the activities of bacteria. It is the work of bacteria in this direction which we call fermentation and putrefaction. Without that work, as we understand it, the existence of life upon the earth would soon come to an end, and the dead and undecomposed bodies of living things and their products of all kinds would lie about unchanged, as they had fallen.

Bacterium termo is the name formerly given to a supposed species of bacteria which was credited with being the producer of putrefaction. The individuals were represented as being short rods, mostly going in pairs, and actively motile. The term has been abandoned since it appears to have included a number of different species.

CHAPTER III.

DISTRIBUTION OF BACTERIA.

The Bacteria of the Soil.—Bacteria are present in the soil in enormous numbers—100,000 or more in 1 c.c. of virgin soil, according to Flügge. The depths to which they penetrate will depend upon the character of the soil and the character of the life upon it, and whether or not it has been artificially disturbed, as by cultivation. In general, at a depth of 1.25 meters (about four feet) the number will have become very small, and a little deeper the soil will be entirely sterile.

The bacilli of tetanus and malignant edema, and bacillus aerogenes capsulatus are present in the soil of many localities. According to Woodhead, certain savage tribes of Africa and the East Indies use as an arrow-poison soil that is capable of producing tetanus. The bacillus of anthrax may be found in soil which has been infected with this organism.

Most of the bacteria of the soil are harmless or useful saprophytes.[1] The *nitrifying bacteria* described by Winogradsky and by Jordan and Richards belong to the latter class. There occur in soil organisms which have the power of converting ammonia into nitrous acid which forms nitrites, and others which complete the change of nitrites into nitrates. Both varieties are widely distributed. These organisms will not grow on ordinary culture-media, and their cultivation presents great difficulties. Probably a good many bacteria have similar properties to some extent. The

---

[1] See Conn, "Agricultural Bacteriology."

work done by nitrifying bacteria in making nitrates from sewage, manure and the like is indispensable to most plant life. Bacteria have also been credited with the assimilation of free, atmospheric nitrogen, resulting in the addition of a valuable proportion of nitrogen compounds to the soil. This is spoken of as *nitrogen fixation.* Inasmuch as a large part of the excrementitious products of animals containing nitrogen are not retained in the soil, where they may be employed as food by plants, but are washed directly or indirectly into the sea by means of sewage and the rivers, it will be seen that the supply of nitrogen compounds might be in a way to suffer gradual exhaustion. Furthermore, it has already been noticed (page 130) that one of the products of decomposition by bacteria is nitrogen, which is not available to animals and most plants as food. These facts have met with practical recognition by agriculturists in the adoption of various methods of fertilizing the soil. It appears that the roots of peas, beans, clover, alfalfa and some other plants frequently present minute tubercles. These tubercles are pathological growths, caused by the development of microörganisms related to the bacteria. The organisms appear to have the power of assimilating atmospheric nitrogen and of converting it into nitrogen compounds. Experiments show that these observations may be destined to be of great value to the farmer.[1]

The bacteria of the soil may easily be studied in plate-cultures made from small portions of soil collected with the necessary precautions to avoid contamination, or plate-cultures may be made from sterilized water with which a portion of the soil has been properly mixed. Anaërobic bacteria must be cultivated by the special methods adapted to them.

---

[1] For simple experiments to illustrate these phenomena see Buxton, *Journal of Applied Microscopy,* September, 1902.

Bacteria of the Air.—The bacteria of the air will be found for the most part clinging to solid particles in suspension in the shape of dust. As has already been stated, bacteria will not rise from moist surfaces unless forcibly removed, as by agitation or currents of air. Conditions of dryness and wind tend to increase the number of micro-organisms in the air. They are fewer after a fall of rain or snow, and the number is smaller in winter than in summer. The air of cities contains more germs than that of the country. The atmosphere over the sea and at the tops of high mountains is nearly or wholly free from germs. The bacteria which do occur in the air will seldom be pathogenic. Their character will depend upon the character of the dust. It is obvious that dust which consists in part of the dried, pulverized expectoration of cases of pulmonary tuberculosis may contain tubercle bacilli. Anthrax of the lungs sometimes arises in men who handle the wool of sheep that were infected with anthrax (Wool-sorter's disease), and is due to the inhalation of anthrax spores attached to the wool. It is likely that the atmosphere in the immediate vicinity of cases of the exanthematous fevers may contain the organisms, whatever they may be, that cause these diseases.

In a rough way one may obtain some knowledge of the character of the organisms in the air of a given locality by removing the cover of a Petri dish containing sterilized gelatin or agar for a few minutes, replacing it, and allowing the organisms to develop. In most cases a large proportion of the growths that appear will be moulds. Yeasts are also common, and among the bacteria the micrococci are abundant. Chromogenic varieties are likely to be present.

A few studies of this character will show that the number of organisms that are present depends chiefly upon whether the air is quiet or has recently been disturbed by draughts, gusts of wind, or sweeping. These facts are of fundamental importance in laboratory work, where plate-cultures are being studied, if we wish to avoid contamination of the plates. Among various devices that have been proposed for the accurate study of the organisms of the air, the Sedgwick-Tucker aërobioscope is the simplest and most accurate. It consists of a glass tube, one end of which is drawn out so as to be smaller than the other. The small end contains a quantity of fine granulated sugar; both ends are plugged with cotton, and the instrument is

Fig. 47.

Sedgwick-Tucker aërobioscope.

sterilized. A definite quantity of air is to be aspirated through the large end, after removing the cotton, which may be done by means of a suction-pump applied to the other end, or by siphoning water out of a bottle the upper part of which is connected with the end of the aërobioscope by means of a rubber tube. The sugar acts as a filter and sifts out of the air the microörganisms which are contained in it. Liquefied gelatin or agar may be introduced into the large end of the instrument by means of a bent funnel; and, after replacing the cotton, it may mix with the sugar which dissolves. The culture-medium may be spread around the inside of the larger portion of the tube after the manner of an Esmarch roll-tube. The bacteria which were filtered out by the sugar will develop as so many colonies upon the solidified medium.

Bacteria of Water and of Ice.—The water of rivers, lakes and the ocean always contains bacteria. The number of organisms varies greatly in different places and under different conditions. The number of different species found in water is also very large. Ground-water[1] contains few or no bacteria under normal conditions, and is therefore suitable for a source of water-supply, when a sufficient amount is available. The possibility of contamination of the ground-water from unusual or abnormal conditions should always be eliminated before it is taken for drinking-water. Numerous epidemics of typhoid fever have been traced to contamination of wells. The location of wells with reference to privy-vaults and other possible sources of contamination should be chosen with the greatest care.

The ordinary bacteria of water are harmless, as far as is known.[2] Bad odors and tastes in drinking water that is not polluted with putrid material are usually due to minute green plants (algae).[3] The diseases most commonly disseminated by water are typhoid fever and Asiatic cholera, and probably also dysentery. The spirillum of cholera will usually die in natural water (not sterilized water) inside of two or three weeks; the bacillus of typhoid fever will usually die in two or three weeks. Under exceptional circumstances these organisms may perhaps maintain their vitality for a longer period. They appear, however, to be less hardy than the ordinary water bacteria. As we now understand these diseases, the organisms causing them will be present only in a water-supply which has been contaminated by the excreta from a case of the disease. Notwithstanding the rapid death of these organisms in water, they

---

[1] Ground-water is the water which—originally derived from rain or snow—sinks through superficial porous strata, like gravel, and collects on some underlying, impervious bed of clay or rock.

[2] See Fuller and Johnson, "The Classification of Water Bacteria," *Journal of Experimental Medicine*, Vol. IV., p. 609.

[3] "Contamination of Water Supplies by Algæ." G. T. Moore in Yearbook U. S. Department of Agriculture, 1902.

may exist long enough to infect individuals habitually drinking the water. Many epidemics of cholera and typhoid fever have been traced to water polluted with the discharges from cases of these diseases.

By *self-purification* of water is meant the removal through natural processes of contaminating organisms such as might occur from the discharge of sewage into it. It depends upon the sedimentation of the contaminating material, in the form of mud, upon the growth of the ordinary water-plants and protozoa, upon the exhaustion of the food supply by the growth of bacteria themselves, upon the destructive influence of direct sunlight, and the dilution of the matter added with a large volume of water.[1] It is not usually to be relied upon as a means of freeing the water-supply from pathogenic bacteria.

Storage of water. When water is kept in large reservoirs, the solid particles in it, including bacteria, tend to fall to the bottom. The number of bacteria in a water-supply may be considerably reduced in this way.

Filtration.—Filtration on a large scale has been more commonly in use in the cities of Europe than elsewhere, until lately. Similar filtration-plants now exist in several cities of the United States.

Slow Sand Filtration.—The filter consists of successive layers of stones, coarse and fine gravel. The uppermost layers are of fine sand. The whole filter is from 1 to 2 meters thick. The sand should be 60 cm. in thickness. The upper layers may be removed from time to time, the remainder not becoming less than 30 cm. in thickness. The first water coming from the filter is discarded. The actual filtration is done largely by the slimy sediment which collects on the surface of the layer of fine sand. The filter-beds may be several acres in extent, and are protected by arches of brick or stone. They require renewal occasionally. This

[1] See Jordan, *Journal of Experimental Medicine*, Vol. V., p. 271.

kind of filtration has come largely into use since the cholera epidemic of 1892–93, and it appears to be very effective.

Mechanical Filtration.—This method of filtration is also called the American system. It is more rapid than the preceding method and does not require a large area for filter beds. Although sand is required also, filtration is accomplished by a jelly-like layer of aluminum hydroxide. This product is formed by adding to the water a small quantity of alum. The carbonates in the water decompose the alum and produce aluminum hydroxide. It precipitates as a white, flocculent deposit, entangling solid particles, including bacteria, as coffee is cleared with white of egg. Only a trace of alum should appear in the water. This method of filtration has not been tested so extensively as slow sand filtration, but seems likely to prove efficient. With water poor in carbonates, these may have to be added.

Various methods for the purification of water by means of chemicals have been proposed. The use of ozone for this purpose has met with considerable favor.[1]

The filtration of water on a small scale, as is ordinarily done for domestic purposes, is generally entirely useless. The so-called Pasteur filter of unglazed porcelain is effective if it is properly constructed and if the filter-tubes are sterilized by heat frequently (every few days)—conditions which are seldom complied with. Distillation of water, or thorough boiling will usually be the most practical method for sterilizing drinking-water.

Collection of Samples.—Samples from the water-supply of a city may be drawn from the faucet, but the water should first be allowed to run for half an hour or longer. From other sources the supply should be collected in sterilized tubes or bottles, taking care to avoid contamination. Sternberg bulbs (see Fig. 38) will be found useful for

[1] Consult "Disinfection and Disinfectants," Rosenau, 1902.

small samples. These samples should be examined as quickly as possible, for the water bacteria increase rapidly in number after the samples have been collected. When transportation to some distance is unavoidable the samples should be packed in ice.

The **number of bacteria** may be determined by making plates of a definite quantity of the water with gelatin or agar. The amount examined ordinarily is 1 c.c. When the number of bacteria is very large, a smaller quantity must be taken, and it may be necessary to dilute the sample ten times or more with sterilized water. The amount should be measured with a sterilized, graduated pipette. The water is to be mixed with liquefied gelatin or agar in a tube which has been allowed to cool after melting. After thorough mixing, remove the plug, burn the edge of the tube in the flame, hold in a nearly horizontal position until cool, and pour into a sterilized Petri dish. The number of colonies may be counted on the third or fourth day; the later the better, as some forms develop slowly and may not present visible colonies for several days; but the plates are often spoiled after three or four days by the profuse surface growths of certain forms or by the rapid liquefaction of gelatin, if that be used, by other forms. The number of colonies that develop is supposed to represent the number of individual bacteria contained in the quantity measured. That will probably not always be the case, however, as colonies may develop from a clump of bacteria which have not been separated from one another by the mixing process. Abbott has shown that the number of colonies is usually larger on gelatin plates than upon agar plates, and at the room temperature than in the incubator. This observation illustrates the fact that there are doubtless many kinds of bacteria that do not find favorable conditions for development on ordinary culture-media. The reaction of the medium has an important influence upon the development of these water bacteria in plate-cultures.

When the number of colonies is small, there is no difficulty in counting them as they appear in the ordinary Petri dish. When the number is large some kind of mechanical device may be used to assist counting. The Wolffhügel plate is a large square of glass resting in a wooden frame painted black. The glass plate is ruled in squares. It was designed particularly with reference to the form of plate-cultures first made by Koch. The Petri dish, however, may be placed upon the glass plate and the cross lines be used to assist in counting. Lafar, Pakes and Jeffer recommend a surface painted black, ruled with white lines which represent the radii of a circle, which may be still further subdivided by other lines. Many find counting easier when a black surface divided into squares is employed. An ordinary card with a smooth black surface divided into squares by white lines may be placed under a Petri dish and will be found to serve very well. For the mere examination of the colonies no better surface can be devised than the ferrotype plate used by photographers. The examination of the colonies will be easier if a small hand-lens be used. Care must be taken not to mistake air-bubbles or particles of dirt for colonies of bacteria.

In any case, if possible, all the colonies in the plate should be counted. The number contained within several squares may be counted and the average taken; knowing the size of the squares and the area of the plate, the number contained in the whole plate may be calculated. Such estimations, however, are likely to give results very wide of the truth.

The plating may be done by rolling the medium after the manner of Esmarch. When the number of colonies is not large this may serve very well. Counting may be assisted by drawing lines with ink on the outer surface of the test-tube.

It has been said that a water-supply containing no more than 500 bacteria per cubic centimeter is to be regarded as safe, one having between 500 and 1000 is to be looked upon with suspicion, and that where there are more than 1000 to the cubic centimeter the water is unfit for drinking purposes. It is obvious, however, that the character of the bacteria is of prime importance; that pathogenic organisms may occasionally be present, even when the number of bacteria to the cubic centimeter is small. But knowing the number usually found in a good water-supply, any sudden variation above that number is to be looked upon with suspicion. An increase is to be expected when the water has been subjected to unusual agitation from winds or currents.

The detection of pathogenic bacteria in water[1] involves great difficulties, and our knowledge in this direction is very meagre. Koch and several others have reported finding the spirillum of Asiatic cholera in water. The examination of water-supplies for this organism has disclosed the fact that bacteria resembling the organism of cholera in many respects are not uncommon in water. This adds to the difficulty of detecting the cholera germ in water.

The bacillus of typhoid fever has many times been described as occurring in water-supplies suspected of being contaminated with the excreta of cases of the disease. The interpretation of these observations is at present doubtful. It is now known that several forms of bacteria exist which closely resemble the bacillus of typhoid fever, and which would make its recognition in an unknown specimen very difficult.

It will at once be appreciated that the number of cholera and typhoid organisms necessary to contaminate a consid-

---

[1] See also articles in Part IV. on the bacillus of typhoid fever, bacillus coli communis and spirillum of cholera.

erable body of water, and sufficient to cause an outbreak of
the disease among some of the people drinking the water,
may still be so small that many different cubic centimeters
of the water might be studied before a single one of the
specific organisms would be encountered. Anyone who
has examined plates made from samples of water will recognize the difficulty of detecting one or a few colonies of the
bacteria of cholera or typhoid fever among a hundred or
more colonies of ordinary water-bacteria. The existence of
contamination with animal excreta might, however, be indicated by finding the bacillus coli communis, whose detection
offers a greater prospect of success. It is not certain just
how much importance is to be attributed to the presence of
small numbers of the colon bacillus in water.[1] Until our
knowledge is more complete any suspicious water should be
discarded.

Certain devices have been adopted to hasten the development of the desired bacteria and to retard that of the ordinary water-bacteria. Among these may be mentioned the
influence of the heat of the incubator, which will hasten the
growth of organisms derived from the human body, and
which retards the growth of water-bacteria. Another is
the addition of a solution of peptone to a large quantity of
the water to be examined with a view to assisting the
development of the desired bacteria by furnishing them
suitable food for growth. In another method (Parietti's)
small quantities of carbolic acid are added to bouillon and
mixed with the water, with a view to retarding the development of all except typhoid and colon bacilli. Suspected bacteria may be tested by inoculation into animals; the pos-
session of pathogenic properties creates a probability in
favor of their having come from some contamination with
animal excreta.

[1] Jordan, *Journal of Hygiene*, Vol. I., 1901. Savage, *Journal of Hygiene*, Vol. II., 1902. Winslow & Hunnewell, *Journal Medical Research*,
Vol. VIII., 1902.

Detection of Bacillus coli communis in Water.—To each of a number of fermentation-tubes containing 1 per cent. dextrose-bouillon add some of the suspected water (.1 to 1 c.c.). Place in the incubator. Each day mark the amount of gas that has formed in the closed arm. In three days B. coli communis should render the bouillon strongly acid and produce 50 per cent. of gas (or about that amount). From tubes showing these characters, plates may be made and the usual tests for the colon bacillus applied.[1] (See Part IV.)

Ice.—The bacteriological examination of ice differs in no respect from that of water. Although development may be arrested, the vitality of bacteria is not necessarily impaired by freezing. Prudden found the bacillus of typhoid fever alive in ice after more than one hundred days. However, Sedgwick and Winslow have stated that when typhoid bacilli are frozen in water the majority of them are destroyed. Nevertheless, it is safest to have the source from which ice is taken as carefully scrutinized as that of the water-supply, especially in view of the universal habit of cooling water in the summer time by adding ice directly to the water. It is better to cool water and articles of food by surrounding the vessels containing them with ice.[2]

Bacteria of Milk and Other Foods.[3]—Of the different food substances, milk is probably the most important from a bacteriological point of view. In the first place, most other foods are cooked before eating. Furthermore, cow's milk constitutes a large part of the food of many young infants who are highly susceptible to certain bacteria, or to substances in the milk itself, after it has undergone certain alterations due to bacteria. The milk of the healthy cow as it is secreted in the mammary gland is sterile; however, after milking the cow a little milk generally remains

---

[1] T. Smith, *American Journal Medical Sciences*, Vol. 110, 1895.

[2] Clark, "Bacterial Purification of Water by Freezing," Reports American Public Health Association, Vol. XXVII., 1901.

[3] See Conn, "Bacteria in Milk and its Products," 1903. Russell, "Dairy Bacteriology."

in the milk-ducts and in the lower part of the teat in which numerous bacteria will have developed before the next milking-time.[1] The first milk obtained at a milking should therefore be discarded, as it may contain an excessive number of bacteria.

Contamination with bacteria may occur from the outer surface of the udder of the cow, the hands of the milker or dirty pails, or through agitation of the air of the stable, and in other ways readily conceived of. Bacillus coli communis is often found in milk. Excluding the tubercle bacillus, the organisms which contaminate milk will be pathogenic only in exceptional cases. Occasionally typhoid fever, cholera, and possibly scarlet fever, diphtheria and other diseases are disseminated by means of contaminated milk. In the case of typhoid fever, it is probable that the milk cans have been washed with polluted water; after the cans were filled, a few typhoid bacilli left in drops of water in the cans, might multiply enormously. Streptococci have been found quite frequently in the milk sold in cities.[2] The mixture with the milk of non-pathogenic organisms from the air, and their growth, may induce changes in it which render it unfit for consumption, and even poisonous. These alterations may be evident to the senses, as the ordinary lactic acid fermentation (souring of milk), or they may not. The character of the alterations doubtless varies much with the temperature and with the character of the contaminating bacteria. Summer temperatures of course favor decomposition and fermentation. Specialists in children's diseases attribute to alterations in milk with the formation of poisonous substances a preëminent influence in the production of the intestinal disorders of infancy so common in the summer.

[1] See Harrison and Cumming, "The Bacterial Flora of Freshly Drawn Milk," *Journal of Applied Microscopy,* November, 1902.

[2] See Reed and Ward, "The Significance of the Presence of Streptococci in Market Milk," *American Medicine,* February 14, 1903.

Poisoning with milk, ice-cream or cheese is not rare, as is well known. There are many records of whole companies of individuals having been taken violently ill after having eaten one of these foods from the same source of supply. The symptoms in such cases resemble those produced by irritant mineral poisons such as arsenic: nausea and vomiting, vertigo, dryness of the mouth, sense of burning and constriction in the throat, difficulty in swallowing, cramps and griping pain in the bowels, constipation or diarrhea, general prostration or even collapse. Vaughan isolated from poisonous cheese a ptomaine which he called tyrotoxicon. It appears, however, that other toxins may be present in cheese, and that tyrotoxicon is a somewhat rare poison. Vaughan believes that bacteria of the colon group play an important part in producing poisons in milk and cheese.

To prevent the alteration by bacteria of milk intended to be the food of infants, the practice of sterilizing milk has been largely in vogue. Unfortunately, during sterilization the milk undergoes some kind of alteration which makes it disagree with certain infants. Furthermore, among the organisms which would be likely to contaminate milk the bacilli of hay and potato, whose spores are so excessively resistant, would be prominent, and they are not killed by any process to which the milk intended for an infant's consumption could possibly be subjected in the household. Least of all can sterilization be expected to purify milk in which bacterial poisons are already formed.

The process called pasteurization is designed, not to sterilize the milk completely, but to destroy the vegetative forms of bacteria, and to destroy the ordinary pathogenic bacteria with which the milk might possibly be contaminated.[1] The milk is subjected to a temperature of only

[1] See Journal of Experimental Medicine, Vol. IV., p. 217. "The Thermal Death-point of Tubercle Bacilli in Milk," etc., by T. Smith.

about 70° to 75° C. This temperature is less likely to produce alteration in the milk than sterilization by steam at 100° C. According to Freeman, milk which had been pasteurized at 75° C. and distributed among the poor people of New York City, whose homes were not supplied with ice, usually kept very well even in the summer time (see p. 67).

The number of bacteria in milk may be reduced considerably by the use of the centrifuge (separator).

It has been undertaken recently to do away as far as possible with the contamination usually inevitable in the barnyard and stable by the use of extraordinary measures to keep the cows, and especially their udders, clean; also the hands of the milker and the milk-pails; and by sprinkling the floor of the milk-room to prevent dust.[1] The milk is to be transported to the city on ice. Milk which has been collected in this manner is furnished in several cities in the United States. The cattle from which the milk is derived are regularly inspected by veterinary surgeons as well as subjected periodically to the tuberculin test. The surroundings and drainage of the stables are investigated by physicians and sanitary engineers. The milk is also regularly analyzed by a chemist. It has been found possible to reduce the number of bacteria in milk very noticeably. This milk is of course sold at a considerably higher price than ordinary milk.

The number of bacteria which occur in samples of milk varies greatly. In ordinary milk as furnished by milkmen the number of bacteria to the cubic centimeter is usually many thousands to millions; grocer's milk may contain hundreds of thousands to millions of bacteria to the cubic centimeter; frequently figures are reached which are beyond computation.

Human milk often contains the staphylococcus epider-

[1] See W. H. Park, *Journal of Hygiene*, Vol. I., 1901.

13

midis albus, and not seldom the staphylococcus pyogenes aureus, under normal conditions.

Of the different pathogenic bacteria liable to furnish a source of danger in milk, the most important is the bacillus tuberculosis. Tuberculosis is a disease to which cattle are exceedingly prone. There is good reason to believe that infants acquire tuberculosis through taking as food the milk of tuberculous cows, although the danger from this source has probably been overestimated. The milk of tuberculous cows may contain tubercle bacilli when there is no tuberculous disease of the udder.[1] The frequency of tuberculosis among milch cows sometimes becomes as high as 25 per cent., or even more. Butter derived from the milk of such cows may contain tubercle bacilli. The proper manner for the States to deal with this problem, for it is one that doubtless will fall to the individual States, has not yet been determined. The cost of killing such a large number of valuable cows would be very great. Furthermore, it is by no means certain that this procedure would eradicate the disease. The flesh of cattle also is capable of transmitting tuberculosis, but is a smaller source of danger on account of the universal practice in the United States of thoroughly cooking beef.

" Ripening " of cream and cheese is due to the growth of bacteria which produce agreeable flavors in the butter and cheese. Molds are also important in the ripening of some kinds of cheese.[2]

In examining milk for bacteria the number may be estimated by precisely the same technique as is used for the estimation of bacteria in water, except that the milk must be diluted; otherwise the plates are rendered opaque by

[1] See Mohler, "Infectiveness of Milk of Cows which have Reacted to the Tuberculin Test," U. S. Dept. Agriculture, Bureau Animal Industry, Bull. No. 44, 1903.

[2] Conn, "Agricultural Bacteriology."

the fat. It may be diluted one hundred times with sterilized water; when the number of bacteria is very great a second dilution may be required. Estimations based upon such high dilutions can only be approximate. The quantity taken for examination may be o.1 to 1 c.c. Plates should be made immediately upon collection of the sample. If the milk stands for a few hours at the room temperature in the laboratory, the number of bacteria will become enormously increased.

The detection of a particular kind of pathogenic bacteria in milk or butter involves very great difficulties. Staining of bacteria in milk may be done by the usual methods, but the results are rendered unsatisfactory by the oil in the milk. The demonstration of tubercle bacilli by staining methods is likely to involve many difficulties. In this connection it is necessary to remember the group of bacilli which resemble the tubercle bacillus in resisting decolorization with acids after staining. (See p. 44.) The procedure of injecting milk into guinea-pigs has been resorted to largely, but the results are only obtained after the lapse of weeks, when the development of tuberculosis in the guinea-pigs would indicate that the milk was tuberculous, provided that control guinea-pigs remained healthy. Furthermore, the other acid-proof bacilli which may occur in milk or butter are capable of producing nodules resembling tubercles.[1] (See Bacillus tuberculosis, Part IV.) The most satisfactory plan will be to apply the tuberculin test to the cow from which the milk was derived.

Among the *other articles of food,* those are to be most carefully scrutinized which are to be eaten after little or no cooking, particularly salads, green vegetables, fruits, and the like. Under exceptional circumstances they may become agents for conveying infectious diseases. Conn

---

[1] Rabinowitsch, *Zeitschrift f. Hygiene,* Bd. XXXVII., p. 439.

showed that there was good reason for attributing an epidemic of typhoid fever among students at Middletown, Connecticut, to raw oysters. After having been collected from the oyster-beds, these oysters were placed in a small stream to fatten, where they were exposed to contamination from a sewer. Into this sewer the discharges of a case of typhoid fever were found to have been running at the time when the oysters were fattening. An epidemic at Atlantic City, New Jersey, in 1902, was traced to nearly similar causes and conditions.[1]

The ordinary processes for curing and salting meat cannot be relied upon to destroy pathogenic bacteria.

Cases of poisoning by eating oysters, fish, meat in the form of sausage or canned meat, and other articles of food are not rare. These cases belong to the same class as those poisoned by milk and cheese already mentioned. They are due to products of bacterial decomposition. Such affections are quite commonly called " ptomaine poisoning," although the poisons are not ptomaines in most cases. Probably a number of bacteria exist which are capable of affecting changes in meat and other foods either before or after ingestion. Among these are an anaërobic bacillus described by Van Ermengem (B. botulinus), bacillus enteridis (Gaertner) and members of the groups of which B. coli communis and B. proteus are types.[2]

---

[1] *Philadelphia Medical Journal*, November 1, 1902.

[2] See Vaughan and Novy, "The Cellular Toxins," 1902. Ohlmacher, " Food-Intoxication from Oatmeal," *Journal of Medical Research*, Vol. VII., p. 420. Galeotti and Zardo, *Centralblatt f. Bakteriologie*, Vol. XXAI., 1902, Orig. p. 593.

CHAPTER IV.

THE BACTERIA OF THE NORMAL HUMAN BODY.

THE numerous solid tissues and organs of the human body, the fluids circulating in the interior like the blood and lymph, and the cavities that have no connection with the outer world, are entirely free from bacteria.[1]   So also the maxillary, ethmoidal and frontal sinuses, middle ear,[2] urinary bladder, uterus and Fallopian tubes, and to a less extent the lungs and gall-bladder,[3] although having external connections, are usually sterile when in a healthy condition. When bacteria do enter the tissues from any of the surfaces their progress is checked by means of the activities of the cells or fluids of the body, and if they succeed in penetrating to any considerable distance their advance is usually arrested by the nearest group of lymph-nodes, which appear to be important safeguards for preventing the dissemination of bacteria throughout the body.   As a rule, the secretions of the mucous membranes are inimical to bacteria.

The skin, as might be expected, is liable to have upon it numerous bacteria, especially micrococci, and moulds.

[1] This view is not upheld by the experiments of Ford, who found small numbers of bacteria in the normal organs of rabbits, cats and dogs in the majority of those examined.   The species of bacteria obtained were mostly common saprophytes, and to some extent constant in the same kind of animal. *Journal of Hygiene*, Vol. I., 1901.

[2] Calamida and Bertarelli, *Centralblatt f. Bakteriologie*, Vol. XXXII., 1902, Orig. p. 428.   Törne, *Ibidem*, XXXIII., 1903, p. 250.   Hasslauer, *Ibidem*, Ref. XXXII., p. 174.   An examination of these articles will show that investigators disagree somewhat, with regard to the sterility of these cavities.

[3] See Review on the "Bacteriology of the Gall-Bladder and its Ducts," *American Journal Medical Sciences*, Vol. 123, p. 372.

The staphylococcus pyogenes aureus, the streptococcus pyogenes, the bacillus pyocyaneus and the bacillus coli communis sometimes occur on the skin. According to Welch, it always contains the staphylococcus epidermidis albus, which may be a form of the staphylococcus pyogenes albus. This organism is of some importance to surgeons on account of its relation to the cleansing of the skin before operations. It seems impossible, by any amount of cleaning, to dislodge all of the germs in the skin, especially those under the nails.

The bacteria of the exposed mucous membranes like the conjunctiva and the nasal cavity[1] and the mouth cavity will naturally be very fluctuating both in quantity and quality; they will be, in fact, those which happen to fall upon the surface or to be drawn in from the external air.

In the mouth, however, there is a certain group of organisms more or less characteristic of it, many of which have not been successfully cultivated. These have been thoroughly studied by Miller, to whose works students are referred.[2]

Several species of spirilla have been discovered in the mouth and are found along the margins of the gums. The leptothrix buccalis, and related organisms which have a long, ribbon-like form, also occur in the mouth. · The micrococcus lanceolatus (or pneumococcus) appears to be present in many human mouths. In 15 to 20 per cent. of human mouths this organism is sufficiently virulent to produce fatal septicemia when inoculated into susceptible animals. Pyogenic bacteria, especially streptococci, occur frequently, although not regularly, in the mouth. Putrefactive

[1] Hasslauer, "Die Bakterien flora der gesunden und kranken Nasenschleimhaut," *Centralblatt f. Bakteriologie,* Vol. XXXIII., 1902, Orig. p. 47.

[2] Miller, "Microörganisms of the Mouth." For a recent review on the bacteria of the mouth, see Madzar, *Centralblatt f. Bakteriologie,* Vol. XXXI., 1902, Ref. p. 489; Vol. XXXII., p. 609.

bacteria acting on particles of food about the teeth produce the bad odor from the mouths of persons of careless habits. According to Miller, bacteria play an important part in the production of dental caries. Certain of the bacteria of the mouth produce fermentation in the vicinity of the teeth with the formation of acids, which dissolve the calcium salts of the teeth. The softening and destruction of the decalcified matrix is then accomplished by other and liquefying forms.

The expired air coming from the mouth and nose, contrary to the popular notion, is free from bacteria, excepting those which become forcibly detached, as by efforts of sneezing and coughing.

Among the other exposed mucous surfaces, the urinary meatus and the vagina may be included. The urinary meatus and at least part of the urethra will be found to contain bacteria, which, in health, should be non-pathogenic, although interest attaches to the fact that diplococci have been described which behaved with stains in the same manner as the gonococcus (pseudo-gonococci).

There has been much dispute as to whether or not the pyogenic bacteria occur in the vagina normally. It is probable that the healthy vagina is in most cases free from the pyogenic bacteria; although bacteria of some sort are always present, and the pyogenic bacteria may exceptionally be found there in health. The normal secretion of the vagina has a bactericidal influence which may be attributed in part to its acidity. The upper part of the normal cervix uteri is sterile, while bacteria are present in the lower part.

According to Döderlein the properties of the vaginal secretion are due to bacilli which very commonly occur in it. The secretion is most abundant and important during pregnancy.[1]

[1] J. W. Williams, "Obstetrics, A Text-Book, etc.," 1903, pp. 34, 773-775. Wadsworth, *American Journal of Obstetrics*, Vol. XLIII., 1901.

The smegma of the external genitals contains numerous bacteria, among which are frequently found bacilli which retain their color after treatment with acids in the Gabbett method for staining tubercle bacilli. It is uncertain whether these bacilli form a special group of organisms by themselves, having as one of their properties the power of retaining the stain after acids, or whether they are bacilli of no particular sort, which resist acids after staining owing to the oily material with which they have been impregnated in this peculiar secretion. These organisms must be taken into account in staining for tubercle bacilli, urine or other secretions which might accidentally contain particles of smegma. Usually the employment of alcohol after the acid will remove the color from the smegma bacilli (Hueppe). Sometimes smegma bacilli are as resistant as tubercle bacilli to decolorizing agents (Welch); see page 44. Similar acid-proof bacilli occur about the genitals of the domestic animals.[1]

The bacteria of the stomach and intestines are of great interest and importance. The alimentary tract of new-born infants and the meconium are sterile. In from four to eighteen hours organisms begin to appear. They may enter either from the mouth or the anus. There seems to be no constancy in the nature of the forms which are found at first, but their character depends upon the surroundings.

The bacterial inhabitants of the stomach are less constant than we shall find those of the intestines to be. Under normal circumstances they seem to be those introduced from the mouth. Different investigators, at all events, have met with quite different species. It appears that the hydrochloric acid (about 2 parts per thousand) present in the gastric juice at the height of digestion possesses decided germicidal properties. This germicidal power exercises a restraining influence upon fermentation due to bacteria, and

[1] Cowie, *Journal Experimental Medicine,* Vol. V., p. 205.

probably serves as a safeguard against the introduction of pathogenic germs into the intestines. That is particularly important in the case of the spirillum of cholera, which is excessively sensitive to the action of acids. Nevertheless many bacteria are able to reach the intestines uninjured, as the acidity of the gastric juice does not reach its height until some hours after eating. Such bacteria will be those which are most resistant and those which form spores. In the intervals when hydrochloric acid is absent from the stomach, lactic acid appears. It is formed from carbohydrates by a large number of species of bacteria. In conditions of fermentation, sarcina ventriculi and yeasts may be present in large numbers; in the healthy stomach they occur in much smaller numbers.

The intestine of the infant in whom feeding has become well established was found by Escherich to contain two principal species of bacteria—in the lower part of the intestine the bacillus coli communis, in the upper part the bacillus lactis aërogenes. More recently it has been shown that the stools of milk-fed infants, and to a less extent of adults, contain large numbers of anaërobic bacilli, which stain by Gram's method (bacillus bifidus—Tissier, bacillus acidophilus—Moro). These bacteria have not been fully studied.[1]

The number of bacteria in a milligram of human fecal matter has been estimated at from seventy thousand to thirty-three million. It is estimated that about one-third of the fecal matter of adults consists of bacteria. The small intestine of adults has been found by different observers to contain very different species. The majority of these appear to have been introduced from the mouth in food or water. The bacillus coli communis, however, occurs invariably in health not only in the intestine of man, but also

[1] Metchnikoff, "Les Microbes Intestinaux," *Bulletin de l'Institut Pasteur,* May 15 and 30, 1903.

in that of many animals, especially in the lower part.[1]   The pyogenic micrococci very often occur in the intestine.

In the case of ruminant animals like the cow and sheep, the decomposition of cellulose, which forms so large a part of their food, appears to be effected by bacteria. Bacteria having this power are constantly found in the stomachs of ruminants. The best known species is that called bacillus amylobacter. It is questionable whether the products of the decomposition of cellulose have any nutritive value.

Pasteur some years ago expressed the opinion that if animals could be placed in such surroundings that bacteria could be excluded from the alimentary canal and the food, life would be impossible. This view has excited much controversy, and was apparently disproved by the experiments of Nuttall and Thierfelder. These investigators succeeded in removing guinea-pigs from the mother by Cæsarean section, and in keeping them alive in sterile surroundings, upon sterile food, so that the contents of the alimentary canal remained sterile. Schottelius, who worked with chickens, obtained contrary results, however, so that this interesting question is still undecided.

[1] Moore and Wright, "Bacillus coli communis from Certain Species of Domesticated Animals," *American Medicine*, March, 1902.

CHAPTER V.

BACTERIA IN DISEASE.

To the physician and the student of medicine the study of bacteriology is interesting chiefly on account of the great importance attributed to bacteria in producing disease. The presence in an organism of one or a number of organisms of another species, which flourish as parasites upon the first, is a phenomenon of very wide occurrence in nature. It is, in fact, nearly universal. It may be observed among plants as well as animals, for example in the familiar galls seen on some of the higher plants, and mostly caused by the larvæ of insects harbored by the plant. We also find animals, such as tape-worms and the trichina spiralis, living as parasites upon other animals. The functions of the bacteria make them peculiarly suited to leading a parasitic existence. The fact that they possess no chlorophyll, and that they are therefore unable to form carbon compounds from the carbon dioxide of the atmosphere, makes it necessary for them to secure such compounds from pre-existing organic matter. Many of them, furthermore, flourish better when they are able to obtain nitrogenous food from organic matter rather than from inorganic salts containing nitrogen. Most bacteria find the necessary nutriment in the dead bodies of other animals and plants; they constitute what are known as saprophytes. But some of them flourish upon the living bodies of other plants and animals in whom they may produce disease.

The phenomena of disease, we shall find, are very largely due to the numerous waste products of the activities of bacteria, which act as poisons to the host.

The diseases of plants known to be caused by bacteria are not very numerous. Among them may be mentioned pear-blight, due to micrococcus amylovorus.[1] Among lower animals bacteria very frequently produce diseases—for example, chicken-cholera, symptomatic anthrax, erysipelas of swine, hog-cholera, tuberculosis, anthrax and glanders.

Koch formulated certain rules which he considered must be complied with in order to prove that any microörganism was the cause of a particular disease:

First. That the organism should always be found microscopically in the bodies of animals having the disease; that it should be found in that disease and no other; that it should occur in such numbers and be distributed in such a manner as to explain the lesions of the disease.

Second. That the organism should be obtained from the diseased animal and propagated in pure culture outside of the body.

Third. That the inoculation of these germs in pure cultures, which had been freed by successive transplantations from the smallest particle of matter taken from the original animal, should produce the same disease in a susceptible animal.

Fourth. That the organism should be found in the lesions thus produced in the animal.

An *infectious disease* is a disease which is caused by a microörganism growing in the body of the animal having the disease. Such microörganisms are usually bacteria, but not always; for example, malaria is produced by a minute animal organism.

A *contagious disease* is one which is acquired from an individual having the disease. Most contagious diseases are infectious, but infectious diseases are not necessarily contagious. The words are often used very loosely, and

---

[1] See E. Smith in *Centralblatt f. Bakteriologie,* etc., Zweite Abtheilung, Bd. V., p. 271; Bd. VII., p. 88.

it is no longer possible or very desirable to draw the line sharply between them. *Fomites* are the materials on which the contagious element is conveyed.

A *miasmatic disease* is a variety of infection in which the microörganisms are not received from another case of the disease, but are derived from the external world, particularly through foul air.

The following is a list of the most important diseases of man caused by bacteria. The proof as required by the rules of Koch is not complete for all of them:

Tuberculosis,	Gonorrhea,
Leprosy,	Chancroid or soft chancre,
Glanders,	Lobar pneumonia,
Anthrax,	Influenza,
Tetanus,	Diphtheria,
Malignant edema,	Typhoid fever,
Bubonic plague,	Dysentery (not amœbic dysentery),
Malta Fever,	
Suppuration and certain inflammatory conditions allied to it,	Asiatic cholera,
	Relapsing fever,
	Rhinoscleroma (?),
Erysipelas,	Actinomycosis.

Malaria and amœbic dysentery are caused by microscopic animal organisms (protozoa). It has been claimed that small-pox is caused by protozoa; this view has acquired added interest from the recent researches of Councilman. Recent work indicates that the " sleeping sickness " (of Africa) and yellow fever may be caused by protozoa (see appendix). Thrush and certain parasitic skin diseases are caused by fungi of more highly organized structure than bacteria.

In each of the following diseases there is good reason to think that the cause is some kind of microörganism, but it has not yet been discovered.

Syphilis,	Mumps,
Chicken-pox,	Whooping-cough,
Measles,	Yellow fever,
Scarlet fever,	Typhus fever,
German measles,	Hydrophobia,
	Dengue.

Rheumatic fever and Beri-beri would be placed in this list by many writers.

The causes of these diseases have been very carefully sought for by ordinary bacteriological methods with indecisive results. Some of them no doubt are due to bacteria. In recent years numerous observers have described a diplococcus or short streptococcus as the cause of rheumatic fever or acute rheumatism. In the case of yellow fever Sanarelli described an organism (bacillus icteroides) as its cause, but his view is not upheld by most of those who have worked on yellow fever.[1] The bacillus described by a number of observers as having been found in cases of whooping-cough may also be the cause of that disease.[2] Bacilli have also been described in cases of measles on several occasions. Lustgarten has described a bacillus found in the lesions of syphilis which resembles tubercle and smegma bacilli. More recently Joseph and Piorkowsky[3] have cultivated another bacillus from cases of syphilis; how much importance should be attached to it cannot as yet be stated. It is likely that for some of the diseases mentioned other procedures than the usual methods of research will have to be devised in order that the cause may be discovered. The protozoa may play a part in the etiology of some of them. Roux believes

[1] Sanarelli, *La Semaine Médicale*, April 4, 1900. Reed and Carroll, *Journal Experimental Medicine*, Vol. V.

[2] See Czaplewski, *Centralblatt f. Bakteriologie*, Bd. XXIV., 1898, p. 865.

[3] *Centralblatt f. Bakteriologie*, Vol. XXXI., 1902, Orig. p. 445. *Berliner klin. Wochenschrift*, 1902, Nos. 13 and 14.

that contagious pleuro-pneumonia of cattle is due to a microbe so minute that it is barely visible with the highest powers of the microscope, so that its outlines and its morphology can not be studied. The virus of this disease remains virulent after being passed through a Pasteur filter, showing that it is small enough to go through its pores. Similar experiments have succeeded with a number of other affections of animals (of which the best known is foot and mouth disease). The virus may pass through a Pasteur or Berkenfeld filter of a certain coarseness, but is restrained by one sufficiently fine. The most important of the diseases in this class is yellow fever. Reed and Carroll found that the infective agent of yellow fever is in the blood, and that the serum could produce yellow fever in a non-immune person after filtration through a Berkenfeld filter.[1] These facts suggest the possibility that failure to find the causes of some other diseases may lie in the fact that their organisms are so small as to be nearly or entirely invisible to the microscope.

Modes of Introduction.—There are various avenues by which bacteria may enter the body to produce disease. Infection of the embryo through the ovum or semen seems to be of rare occurrence. Syphilis (which may not be due to bacteria) is transmitted in this manner. The embryo may be infected through the placenta, although not commonly. The bacilli of typhoid fever and the pus-forming bacteria have been known to be conveyed through it. Tuberculosis may also be transmitted through the placenta, how frequently is still uncertain. The comparatively common occurrence of endocarditis on the right side of the heart in the fetus may be due to placental infection. Oc-

---

[1] See Reed and Carroll, *American Medicine*, February 22, 1902. For an admirable review of this subject see Roux, "Sur les Microbes dits 'Invisibles,'" *Bulletin de l'Institut Pasteur*, Vol. I., Nos. 1 and 2.

casionally the exanthematous fevers are transmitted from the mother to the fetus.

The surfaces covered with thick stratified epithelium are not likely to be penetrated by bacteria excepting by direct introduction through some wound or other lesion. This, for instance, is true of the skin, the mouth, the vagina and bladder. The infection of bubonic plague appears to be introduced most often by means of wounds in the skin. Bacteria more easily penetrate surfaces having a thin columnar epithelium such as occurs in the intestines, the middle ear, bronchi and bronchial tubes, uterus and Fallopian tubes.

The thin, flat epithelial cells of the air-vesicles of the lungs, as would be expected, seem to be passed with comparative ease. On epithelial surfaces covered with cilia, as in the bronchi and bronchial tubes, the Eustachian tubes, the uterus and Fallopian tubes, the current toward the exterior created by the cilia acts beneficially in removing bacteria.

The tonsils and lymph-follicles of the intestines, especially the lymphoid tissue of the ileum and the vermiform appendix, are points where bacterial invasion frequently begins. The lymphoid tissue of the appendix may have some influence in predisposing to infection at that point and to appendicitis. On the other hand, it is certain that the progress of many infections is checked by the lymph-nodes. That is repeatedly seen in the ordinary post-mortem wound where the spread of the inflammation along the arm is checked suddenly at the elbow or axilla. The participation of the lymphoid structures in most infections is well known. How far this is a conservative process it is impossible to say.

In most cases of infectious disease a point of entrance for the bacteria may be discovered. As a rule, the invading microbes produce a lesion at the point where they are introduced, as in the familiar cases of boils and carbuncles when pyogenic bacteria enter the skin, or of the tubercles found in the lungs when the bacilli lodge in the respiratory tract. However, there are cases of septicemia and pyemia in which the most careful search fails to reveal the place at which the bacteria entered. The bacilli of plague usually produce no reaction at the point of entrance.

It is probable that tubercle bacilli may pass through thin epithelial surfaces and lodge in the deeper structures underneath, where they produce definite lesions. For example, they may pass by the lungs and enter the bronchial glands, and form tubercles in that situation.

Experiments on animals have shown that bacteria may be very rapidly disseminated after their introduction. The inoculation of mice, for instance, with anthrax bacilli has been known to prove fatal, although the wound was washed immediately with the strongest antiseptic solutions or the part amputated within a few minutes.

The manner in which infectious agents reach human beings varies considerably. Generally speaking, the most important element will be found to be direct or indirect connection with another case of the same disease. W. H. Park was able to cultivate diphtheria bacilli from bed clothing soiled by the expectoration of diphtheria cases. Baldwin has shown that tubercle bacilli may be found on the hands of patients having pulmonary tuberculosis, especially those who expectorate on handkerchiefs.

Excepting under certain special conditions, the **Air** will not contain the germs of disease. The dried pulverized sputum of cases of pulmonary tuberculosis may float in the atmosphere as dust which will contain tubercle bacilli. Flügge states that powerful expiratory efforts like coughing and sneezing may carry tubercle bacilli with small particles of secretion into the air, and that they may remain in suspension some time. The pus-producing bacteria may be present in dust. Infectious elements which happen to be present in the air will usually be attached to particles of dust. Wool-sorter's disease is a name sometimes applied to anthrax in man when acquired by those engaged in the work of handling wool, and in which the anthrax bacilli or spores may be conveyed to the lungs in dust.

The atmosphere in the vicinity of cases of the exanthematous fevers at times probably contains the germs of these diseases.

Water is the usual medium for the transmission of the infection in typhoid fever, and Asiatic cholera, and probably all forms of dysentery.

Milk from tuberculous cows may carry the bacilli of tuberculosis; it is of most importance in the case of young infants. Typhoid fever and cholera, and probably scarlet fever and diphtheria are sometimes conveyed through the medium of milk. Bacteria may reach the intestines in *uncooked food*, fruit and vegetables.

The **Soil** is of importance in connection with tetanus and malignant edema, whose bacteria are frequently found in soil. Bacillus aërogenes capsulatus may occur in the soil, and may infect dirty wounds. The spores of anthrax bacilli are present in the soil of certain localities, and may produce anthrax in cattle.

Flies and other insects and related animals are capable of carrying the bacteria of disease. Under suitable conditions flies play an important part in transporting the bacteria of cholera and typhoid fever from the excreta of these diseases to food substances, which they may contaminate. To what extent diseases are disseminated by fleas, bedbugs and similar creatures is still uncertain.[1]

[1] Nuttall, "Role of Insects, etc., in Disease," Johns Hopkins Hospital Reports, Vol. VIII., 1900.

In this connection it is proper to refer to certain diseases due to animal microörganisms. Malaria is conveyed from man to man by mosquitoes of the genus Anopheles, and is probably transmitted exclusively in this manner. The parasite of malaria undergoes part of its cycle of development in man, and another part in the mosquito. Similarly, in Texas Fever, a disease of cattle, it has been shown by T. Smith that the parasite (a protozoön, Piroplasma) passes from cow to cow through the cattle-tick (Boöphilus annulatus or bovis).[1] In surra, a disease chiefly affecting horses, and in the tsetse-fly disease of animals the parasite (a protozoön, Trypanosoma), is transmitted by the bites of flies.[2] It has recently been shown that the infectious agent of yellow fever may be introduced into man by mosquitoes of the genus Stegomyia. Under the administration of the United States Army yellow fever was suppressed in Havana chiefly by measures intended to prevent the disease from being carried by mosquitoes.[3]

Auto-Infection.—It is possible for the bacteria of disease to be derived from the individual's own body—*auto-infection*. The microbes of lobar pneumonia, for instance, flourish in the mouths of a large number of people. The bacillus coli communis, which constantly inhabits the intestines, may invade other organs and exhibit pathogenic properties when the way is opened up for it by other disease processes.

Bodily Conditions that Dispose to Infection.—The development of an infectious disease may be favored by certain *bodily conditions*. Hunger, cold and exhaustion make the body more liable to the inroads of pathogenic bacteria; so also do anemia and chronic diseases. Those suffering from diabetes, as is well known, are especially liable to

---

[1] See V. A. Moore, "Infectious Diseases of Animals," 1902.
[2] Report on Surra, U. S. Bureau Animal Industry, 1902.
[3] Carroll, *Journal American Medical Association*, May 23, 1903.

infection by the pus-producing bacteria and the bacillus tuberculosis.   Dr. Roswell Park believes that prolonged anesthesia makes patients who have undergone operations more liable to surgical infections, and that absorption of bacterial poisons and auto-intoxication due to the products of disordered metabolism of the patient's own cells, predispose to infection.   Some of the above-mentioned conditions can be imitated in laboratory experiments.   Hens in a normal condition are not susceptible to the anthrax bacillus, but Pasteur succeeded in making them contract anthrax by artificially cooling them.   Frogs, on the other hand, which also are resistant to anthrax, may be made susceptible by keeping them at an abnormally high temperature.   Rats were made more susceptible to anthrax by physical exhaustion produced by making them run a treadmill, and pigeons by starvation.

Abbott found " that the normal vital resistance of rabbits to infection by streptococcus pyogenes is markedly diminished through the influence of *alcohol,* when given daily to the stage of acute intoxication."   It was less noticeable for bacillus coli communis, and not observed for staphylococcus pyogenes aureus.   Pigeons and other animals have been made susceptible to anthrax by intoxicating doses of alcohol.

Climate and *altitude* appear to influence the liability to infection with the tubercle bacillus, which occurs less commonly in Colorado and some other elevated regions than in lower and more densely populated districts.

Age.—In general, infants are more susceptible to infections than adults, though apparently nearly exempt from the exanthematous fevers during the early weeks of life. Osteomyelitis is commoner in infants than in adults, as also is tuberculous meningitis.

How much influence is to be ascribed to *individual predisposition* in contracting or warding off infection is uncertain. Welch says, " the fact that some individuals are attacked, and others, apparently equally exposed to the danger of infection, escape, is not always due to any especial predisposition on the part of the former. It may be that the germs hit the one and miss the other, and we would have no more right to say that the former are especially predisposed than to say that those who fall in battle are predisposed to bullets and those who escape are bullet-proof." It is probable that the importance of an *hereditary tendency* to certain infections, notably tuberculosis, has been overrated.

Race.—The influence of racial predisposition is undeniable. For example, it is known that the negro race is much less susceptible to yellow fever than the white race.

Local conditions often have a most important influence in determining the occurrence of infections. In endocarditis the lesion usually occurs along the line of closure of the heart-valves, indicating that the point subjected to the greatest friction is the part of the endocardium most liable to infection. Regions where there is passive hyperemia are more vulnerable, as is seen in hypostatic pneumonia. Localities which have suffered from previous inflammation or irritation are rendered more liable to subsequent infection, as when the bladder or pelvis of the kidney containing a calculus becomes the seat of a suppurative cystitis or pyelitis.

Local conditions become of great importance in surgery. The surgeon can seldom be certain of dealing with a perfectly aseptic wound, and must rely to a large extent upon the power inherent in the fluids and tissues to prevent the development of bacteria. It is important, therefore, to keep the resisting power of the tissues at the highest possible point. Injury of the tissues disposes the part to infection; so do strangulation and necrosis. In operating, it is to be remembered that hyperemic and edematous parts are more likely to become infected; so also are anemic regions. An infarct of the lung which was originally sterile may be infected with bacteria through inhalation, and undergo suppuration or gangrene. The presence of foreign bodies in the tissues disposes to infection. Injection of the staphylococcus pyogenes aureus into a rabbit's tissues is not always followed by suppuration, but if a foreign body, like a piece of sterilized potato, be inserted at the same time, infection is much more likely to occur. When lesions are produced in the internal viscera of animals by cauterization or crushing and bacteria then injected subcutaneously or into the blood, the bacteria lodge in the lesions and multiply.[1]

Amount of Infectious Material.—A large number of bacteria introduced into the body simultaneously will be more likely to produce infection than a small number. This factor is of less importance with organisms whose virulence is very constant than with those of more variable virulence.

Variability in the Virulence of Bacteria.—The occurrence of an infectious disease depends very largely upon the virulence of the bacteria. Any species of pathogenic bacteria may vary in virulence at different times. In some cases the virulence is not easily lost, as with the anthrax bacillus; in others the virulence is maintained in cultures only with difficulty, as in the case of the micrococcus lanceolatus (of pneumonia) and the streptococcus pyogenes. As a rule, the virulence is likely to be diminished in old cultures. It may sometimes be preserved better in the ice-chest than at the room temperature. The virulence of the anthrax bacillus becomes diminished if it is cultivated at 42° C. Exposure to light and to oxygen tends to weaken the virulence; and also cultivation upon unfavorable media, such as those containing a small proportion of carbolic acid or certain other chemical germicides.

[1] Cheesman and Meltzer, *Journal Experimental Medicine,* Vol. III.

In laboratory work the virulence is usually maintained best by inoculating the bacteria from time to time into susceptible animals. Bacteria coming freshly from infected animals are likely to be highly virulent. The virulence may be increased by beginning with an especially sensitive animal like a very young guinea-pig, and progressively inoculating into less sensitive animals. The infection of relatively insusceptible animals may sometimes be produced by the injection of a very large dose of the bacteria. The addition of the toxic products of the bacteria, which may be obtained by using large doses of cultures in bouillon, makes infection more likely. Cultivation on a particular medium may maintain or increase the virulence.

Finally, the combination of two or more kinds of bacteria may produce infection when neither one would do so alone. On the other hand, it is said that the fatal effects of an inoculation of virulent anthrax bacilli into a susceptible animal may be averted if the animal be inoculated with a culture of bacillus pyocyaneus shortly afterward.

Mixed Infection.—It is not uncommon in disease to find two kinds of bacteria associated together, producing a mixed infection. In diphtheria, very frequently, the bacillus of diphtheria is found to be accompanied in the membrane by the streptococcus pyogenes. The course of the diphtheria may be modified in this manner. The term *secondary infection* is rather loosely used. It is sometimes employed to designate an infection occurring in an individual, the resisting power of whose tissues has been weakened by some chronic organic disease. Such an infection is often called a *terminal infection*. Terminal infections are very common in cases of carcinoma, chronic nephritis, arteriosclerosis, and in many other diseases.

Concerning terminal infections Osler says: " It may seem paradoxical, but there is truth in the statement that persons rarely die of the disease with which they suffer. Secondary infections, or, as we are apt to call them in hospital work, terminal infections, carry off many of the incurable cases in the wards."

The term *secondary infection* is also used for the modification of an infectious process which has been in existence for some time, by infection with a second variety of bacteria. That takes place, for instance, in pulmonary tuberculosis, when the invasion of the already tuberculous lungs by the pyogenic micrococci assists in the formation of cavities. In this sense it will be seen that the term secondary infection is used as a name for a variety of mixed infection. In the secondary, mixed and terminal infections, the bacteria which enter secondarily are likely to be of the pus-producing varieties, especially the streptococcus pyogenes.

As to the mechanism which bacteria make use of in order to produce disease, according to our present knowledge, they work chiefly through the poisonous substances formed by them and deposited in the bodies of the persons suffering from the disease. The theory that bacteria have an important influence through the destruction of substances taken by them from the body of the patient for food, is no longer entitled to much weight; neither are we able in most cases to account for the phenomena of disease by any mechanical action on the part of the bodies of bacteria. That such action does occasionally take place may be seen in experimental anthrax in mice, where the blood-capillaries of the liver and kidneys may be completely plugged with masses of anthrax bacilli. The diseases in which the circulating blood is swarming with bacteria are much commoner in the lower animals than in man.

Toxemia.—By toxemia is meant the absorption of poisonous bacterial products from a localized point of invasion, and their dissemination throughout the body by means of the circulation. We see typical toxemias in diphtheria and tetanus. In surgery the term *sapremia* is used to cover a similar condition of affairs when the absorption proceeds from a wound or denuded surface, as may happen in the puerpural uterus.

Septicemia.—In septicemia there is not only absorption of the bacterial poisons, but the bacteria have invaded the living tissues. Bacteriologists usually employ the word septicemia to describe the wide dissemination of bacteria through the body and the presence of a large number of them in the circulating blood. In this sense septicemias are commoner in lower animals than in man; anthrax and infection with micrococcus lanceolatus would be examples. Typical septicemias in man are found in relapsing fever and certain cases of bubonic plague. For pyemia, see the article on Suppuration, Part IV.

The principal agencies in effecting *recovery* from infectious diseases are the presence or formation in the body of substances which destroy bacteria (lysins), the development of new substances which also neutralize their action (antitoxins), and their destruction by the cells of the body (phagocytosis). These phenomena are discussed in the chapter on immunity. A factor of less importance is the elimination of bacteria by the excretory organs. Experiments on animals indicate that bacteria which have been injected into the body do not appear in the urine until they have damaged the structure of the kidney. In typhoid fever, the bacilli of typhoid may occur in the urine in great numbers; the condition of the kidney in the generality of such cases has not thus far been determined. The extent to which the excretory organs act in eliminating bacterial toxins is not yet known. Some bacteria, as has already been stated, may, in the end, produce substances that are inimical to their own growth.

15

CHAPTER VI.

TOXINS.[1]

IT is now generally believed that in most, if not all of the infectious diseases, the principal symptoms and lesions are to be attributed to the action of poisonous substances formed by the bacteria. The part that bacteria play can be understood best by recalling the work of the saprophytes in producing fermentation and putrefaction. It has already been shown that the poisoning that comes from eating decomposed meat, fish, or cheese results from poisons which bacteria have elaborated in the course of their growth. In infectious diseases we suppose the bacteria to grow inside of the body and to form their poisons *in it;* not before their introduction into it, as in these cases of poisoning with spoiled food. If it were possible for the cells of ordinary yeast to grow in the living human body and to produce alcohol from the grape-sugar of the body-fluids, the person so infected might be expected to suffer from alcoholic intoxication as long as the infection lasted. This *impossible* illustration although not entirely accurate may help to make clear what does happen in an infectious disease due to bacteria, where poisons formed in a manner analogous to the formation of alcohol produce intoxications analogous to alcoholic intoxication. Certain infectious diseases exhibit the element of poisoning by bacterial products in an extremely marked manner. In tetanus the local wound may be trifling, and may seem utterly incapable of having given rise to the

---

[1] For a full consideration of this subject see Vaughan and Novy, "The Cellular Toxins," 1902.

violent muscular spasms from which the patient suffers. In diphtheria, although the condition in the throat may be one of severe inflammation, it is, of itself, insufficient to explain the profound prostration and other symptoms of general poisoning which the case manifests.

The first bacterial poisons to be studied thoroughly were those called ptomaines. Observing the poisonous effects which follow the injection into animals of certain ptomaines derived from bacterial cultures, it was suggested that similar ptomaines, formed by the action of bacteria in the living body, might account for the symptoms of many of the infectious diseases. The ptomaines were most readily studied because of the comparative facility with which they could be isolated in a condition of purity, where their exact chemical nature could be determined.

" A *Ptomaine* is an organic chemical compound, basic in character, formed by the action of bacteria on nitrogenous matter." (VAUGHAN and NOVY.)

The peculiar coloring which distinguishes cultures of the bacillus pyocyaneus is due to a ptomaine called pyocyanin and its derivatives.

A group of substances of a similar nature called leucomaines has been discovered, which are formed within the body and not by the action of the bacteria. *Leucomaines* may then be defined as " basic substances which result from tissue metabolism in the body." (VAUGHAN and NOVY.)

Further study has demonstrated, however, that the characteristic features of the infectious diseases are not due to ptomaines. Some of the poisons formed by bacteria have been described as albumens and have given rise to the name *toxalbumen.* It appears, however, that bacterial poisons are not necessarily of an albuminous nature either, and at the present time it seems best to call the bacterial poisons whose chemical nature is uncertain simply *toxins.*

Substances which produce effects in animals similar to the bacterial poisons may be extracted from certain plants, notably abrin, which is derived from the jequirity bean, and ricin, which comes from the castor-oil bean. The venom of poisonous snakes, which is elaborated by the epithelial cells of certain glands, also acts in much the same manner.

Bacterial poisons are of two principal sorts: (1) Sometimes they appear to be formed as excretions from the bacteria. They then occur in solution in liquid cultures, and they may be separated from the bacteria by filtration. The toxins of diphtheria and tetanus are typical examples. (2) The poisons of most bacteria occur chiefly in the bodies of the bacterial cells. Such toxins are called intracellular. Those of typhoid fever and cholera are examples. Precisely how intracellular toxins are set free in the body of an infected animal is not clearly understood. It is, however, known that many of the bacteria in an infected individual are dead and disintegrated. The theories concerning toxins are also considered in the next chapter.

Sometimes the results of the injection of excessively small doses of a toxin are so tremendous as to have given rise to the suggestion that the toxins may be allied to the ferments, like pepsin and trypsin, in their nature.

Owing to the instability of the toxins it has not been possible to isolate them in a state of purity so as to determine their exact chemical character. They have, nevertheless, been obtained in some cases in an extremely concentrated form. Brieger and Cohn obtained a toxin from tetanus bacilli of which .00000005 gram killed a mouse weighing 15 grams. Roux and Yersin obtained a toxin from diphtheria bacilli of which .00005 gram was capable of killing a guinea-pig. These figures indicate a capability for poisoning that is simply inconceivable. Such properties permit bacteria growing in a comparatively limited area to manifest their evil effects at remote parts of the body.

A curious and unexplained effect of some toxins is the production of minute areas of necrosis in certain viscera, as the liver. Such " focal necroses " have been observed to be formed by the poisons of the bacilli of diphtheria, of typhoid fever, and of the micrococcus lanceolatus (of pneumonia), and following the injection of abrin and ricin.

Besides the poisonous substances produced by the bacilli of diphtheria and of tetanus, toxic substances have been obtained from the spirillum of cholera, the bacillus of typhoid fever, the bacillus coli communis, the bacillus of bubonic plague, and from the bacilli of tuberculosis and glanders. The extract from cultures of tubercle bacilli, called tuberculin, and that from glanders bacilli, called mallein, contain toxins produced by these germs, and will be spoken of in connection with the bacteria themselves. Vaughan[1] has succeeded in cultivating anthrax bacilli, colon bacilli, and other bacteria on large surfaces of solid media, so as to secure quantities of the bacterial cells sufficient for extensive chemical tests. The toxin of the colon bacillus proved to be a very stable substance, and resistant to heat. Most toxins become inactive at comparatively low temperatures. (60° to 70° C.)

Prudden and Hodenpyl found that the injection of dead tubercle bacilli was followed by the development of lesions resembling tubercles, which, of course, did not increase in number or become disseminated.

There is good reason on both clinical and experimental grounds to believe that toxic substances are formed by the micrococcus lanceolatus (of pneumonia). The symptoms of a disease as it occurs in man cannot be imitated in any other case as accurately as happens after the injection of the toxins of tetanus and diphtheria in the lower animals.

[1] *American Medicine*, May 18, 1901. *Journal American Medical Association*, March 28, 1903.

CHAPTER VII.

IMMUNITY.

UNDER the title of immunity a number of nearly related subjects may be discussed. It has already been shown that the body is constantly liable to the attacks of pathogenic microbes, which are endeavoring to effect an entrance. Some of the defences of the body have been described. But it has been found that there are other more subtle and at the same time far more powerful weapons, which usually succeed in repelling the invasion. This will answer the question often asked, Why do we not all constantly have infectious diseases? In case bacteria do gain a foothold in the body and produce what we call an infectious disease, we have to consider the means by which the intruders are overcome. The exemption from a second visitation which often follows one attack of an infectious disease also needs to be accounted for.

Certain facts concerning immunity, some of which were observed many years ago, are extremely interesting, but very difficult of explanation. Even in the light of recent bacteriological researches their interpretation is by no means clear. The immunity which an individual who has suffered from an attack of measles or scarlet fever possesses from a second attack of the same disease is well known; so also is the immunity from small-pox which is conferred by vaccination. Such an immunity is called " acquired." There is also a " natural " immunity. Field-mice are susceptible to glanders and house-mice are not. House-mice are susceptible to mouse-septicemia and field-mice are not. Although sheep, as a rule, are easily infected with anthrax, this disease seldom occurs in sheep of the Algerian variety or race. The immunity which belongs to a race, but not to a whole species, is sometimes called " racial."

The occurrence of immunity from a second attack of an infectious disease has given rise to numerous hypotheses in the past. One theory supposed that after an attack of the disease certain bacterial products are retained within the body which prevent a second invasion. Another theory supposed that the attack of the disease exhausts the supply of some substance necessary for the growth of the microbes, as plants sometimes exhaust the soil they grow in.

It will be best to deal first with the results of some of the experimental attempts to produce immunity. The more recent theories will then be considered.

Small-pox and Vaccination.—The origin of vaccination against small-pox with the virus of cow-pox has been described in the historical sketch (p. 21). The nature of the protection furnished by this virus has been the subject of much controversy. The opinion of the present day inclines to regarding vaccinia as small-pox which has been modified by passage through a relatively insusceptible animal. Certainly there are many analogies between the protection against small-pox afforded by vaccination and the other examples of artificial immunity mentioned below.

This question cannot be settled with certainty until the organisms causing small-pox and vaccinia have been isolated in pure culture. Their identity and mode of action may then be determined.

Small-pox has been inoculated into calves and passed through other calves in succession, producing finally an eruption indistinguishable from cow-pox. Lymph taken from such calves has been used successfully to vaccinate children. Not only does cow-pox protect against small-pox, but it has been shown that small-pox protects against cow-pox.

Immunity Produced by Inoculation with Bacteria of Diminished Virulence.—Pasteur conceived the idea of attenuating the virulence of the bacilli of fowl-cholera by prolonged exposure to the air. He made use of the attenuated virus as a vaccine against the disease.

A nearly similar principle was shortly afterward applied by him to the preparation of a vaccine against anthrax. When anthrax bacilli were cultivated at a temperature of 43°C., Pasteur obtained bacilli of very slight virulence. Such bacilli did not produce death when inoculated into animals that were ordinarily susceptible. Yet animals that were vaccinated with this virus were able afterward to resist inoculation with fully virulent anthrax bacilli. (See Bacillus anthracis, Part IV.)

In the case of erysipelas of swine (French *rouget;* German, *Schweinerothlauf*) Pasteur secured bacilli of diminished virulence by injecting virulent bacilli into relatively insusceptible animals. The animal used was the rabbit. The bacilli were passed through several rabbits in succession. Cultures taken from the last of the series produced a milder form of the disease and an amount of immunity, the value of which is in dispute.

In still another disease, black leg of cattle or symptomatic anthrax (French, *charbon symptomatique;* German, *Rauschbrand*), an attenuated virus is secured by the use of heat. The pulp from the infected muscle of a diseased animal, containing the bacilli, is squeezed from it and heated to a temperature of 95° to 99° C. for six hours. The dried material mixed with water constitutes the vaccine. The Department of Agriculture of the United States now furnishes this vaccine free to farmers. The results of this method are said to be very gratifying.[1]

[1] See Recent Annual Reports, Bureau of Animal Industry, U. S. Department of Agriculture.

In the human disease, bubonic plague, a nearly similar procedure has been proposed by Haffkine. To protect against plague, cultures of plague bacilli, previously sterilized by heat and carbolic acid are injected. (See article on Bubonic Plague, Part IV.)

Inoculation Against Rabies or Hydrophobia.—The immunity produced in this case probably depends upon principles similar to those underlying the examples related on the preceding pages. But this question cannot be regarded as settled until the organism of rabies has been isolated and cultivated. Attempts to discover this organism have, as yet, been futile. Pasteur discovered that rabbits are susceptible to rabies when portions of the medulla oblongata of a dog which has died of the disease are inserted beneath the dura mater of the rabbit. Spinal cords taken from rabbits thus injected are placed in a desiccating chamber. Under these circumstances the unknown virus undergoes a diminution in virulence. Emulsions are made from spinal cords desiccated in this manner. First the patient is injected with part of an emulsion from a cord which has been desiccated a longer time, and in which the virulence of the poison has been much reduced. Injections are then made at intervals from cords that have been subjected to desiccation for shorter and shorter periods, and therefore of greater and greater virulence. At the end of about the twenty-fifth day the patient is supposed to be immune from rabies. The period of incubation in rabies is longer than in most infectious diseases, being usually one to two months. Although the injections take place after he has been bitten by a rabid dog, it is hoped the patient may be rendered immune before the period of incubation has ended.[1]

Reports of cases managed according to this method have been conflicting in the past. However, there seems no longer

[1] F. Cabot, "The Dilution Method of Immunization from Rabies," *Journal Experimental Medicine*, Vol. IV., p. 181.

to be any question as to its efficiency. Laboratories where Pasteur's method of treatment is used now exist in all parts of the world. According to statistics collected by Ravenel, based on many thousands of cases, the mortality from rabies in those so treated is less than one per cent.[1]

Antitoxins.—The first efforts to point out the way along which antitoxins might be secured were made by Salmon and Smith in 1886. In their experiments pigeons were injected with filtrates from cultures containing the products resulting from growth of the hog-cholera bacillus. Such pigeons were found to be immune to this bacillus, which is pathogenic to ordinary pigeons.

As was stated in the last chapter, bacterial poisons may be of two sorts. In one group the poisons occur chiefly within the bodies of the bacteria. This group seems to contain the majority of the pathogenic bacteria. Methods of protection against infections caused by them will be considered hereafter. In the other group, the poisons do not, for the most part, remain in the bodies of the bacteria, but are readily diffused from them into their surroundings. It is for the bacteria of the latter group that antitoxins have been made. Its most important members are the bacilli of diphtheria and tetanus. Their poisons may be found in the culture-media in which they have grown. The principle employed in preparing antitoxins was established by Behring. The bacilli

---

[1] In cases of dog-bite there is often some doubt as to whether or not the dog was rabid. The dog should be allowed to die under observation. A large ganglion, as that of the pneumogastric nerve, should be removed and placed in absolute alcohol or ten per cent. formaldehyde solution. Microscopic changes have been observed in the ganglia which seem to be quite constant in rabies. See Ravenel and McCarthy, *University of Pennsylvania Medical Bulletin*, June, 1901; also editorial in *Philadelphia Medical Journal*, March 14, 1903. The diagnosis may be made by inoculating a rabbit as above described, but the results are not obtained quickly enough to be of value. See V. A. Moore, "Infectious Diseases of Animals."

are cultivated in bouillon. The cultures are freed from all living bacilli by filtration. The liquid filtrate contains the toxin. This filtrate is injected into healthy animals, usually horses, in increasing doses. Eventually enormous doses of toxin are given, and the animal acquires a high degree of immunity. The blood of the animal is withdrawn, taking care to avoid contamination. The serum of the blood is collected and constitutes the antitoxin. Such antitoxins have produced brilliant results in the treatment of diphtheria, and have given partial success with tetanus. (See the articles on the bacteria of these diseases.)

Ehrlich discovered that the vegetable toxins, abrin and ricin, behave in a manner very similar to soluble bacterial poisons when injected into animals, and that by their injection an immunity for the same poisons may be secured. There is an analogy between the tolerance acquired in this manner from bacterial and other toxins and that which victims of the morphine and cocaine habits have for immense doses of those drugs. Ehrlich also found that the milk of animals which had been immunized against abrin and ricin might confer immunity upon young animals. In most cases we look to the blood-serum for the immunizing agent.

Active and Passive Immunity.—The kind of immunity which results from the injection of substances from immunized animals is called " passive immunity." Diphtheria and tetanus antitoxins produce passive immunity. " Active immunity " may be brought about in several ways:

(1) By an ordinary attack of an infectious disease; (2) by an attack excited artificially through inoculation with small doses of virulent cultures, or (3) large doses of attenuated cultures; (4) or by the injection of bacterial products (toxins) freed from the bacteria themselves. Pasteur's methods of protective inoculation for anthrax, etc., and Haffkine's injections for bubonic plague produce active

immunity. Active immunity is usually more enduring than passive immunity. Passive immunity, established through the direct introduction of antitoxins, may be brought about more quickly than would be possible for an active immunity.

Theories of Immunity.

Phagocytosis.[1]—Metchnikoff described under the name " phagocytosis " a phenomenon which, he maintained, could explain immunity and recovery from bacterial invasion. This theory is based on the well-known fact that certain cells of the body have the power of surrounding and ingesting foreign substances. The cells in question are chiefly poly-nuclear leucocytes, but to some extent other leucocytes, and endothelial and other cells. There are many examples of this process. The leucocytes of the lungs constantly take up small bits of carbon inhaled with the air. Particles of car-mine injected into the tissues will later be found within leucocytes. After a hemorrhage, phagocytic cells may be found containing red blood-corpuscles or particles of blood pigment. The presumption is that phagocytic cells serve to remove irritating and foreign bodies to less sensitive parts. Metchnikoff showed that phagocytes also absorb bits of de-generating or useless tissue. Such particles disintegrate and their identity is lost. They are digested and become a part of the protoplasm of the phagocytes. This process is seen when the tail of the tadpole shortens. The superfluous part is absorbed, at least in part, by phagocytic leucocytes. Metchnikoff's observations were made largely on the in-vertebrates, whose transparent bodies may be studied while living. One illustration was furnished by a small crustacean (*Daphnia* or water-flea), which was often infected with a fungus. Some infected individuals died, others recovered. Metchnikoff found that the cells of the fungus might be

---

[1] Greek, φαγεῖν, to eat; κύτος, a cell.

ingested and destroyed by the leucocytes of the *Daphnia*. He described the history of this disease as a contest between the parasitic cells and the phagocytes, in which either might succeed. Similarly, when anthrax bacilli were introduced into frogs, which are immune from anthrax, the bacilli were ingested by the frog's leucocytes. Metchnikoff[1] contended that this function of leucocytes and other phagocytic cells constituted the principal defence of the body against bacteria.

Other investigators also have seen bacteria enclosed within the bodies of leucocytes. It has been urged by some that the bacteria are already dead when the leucocytes devour them. In some cases, as with the gonococcus which is commonly found enclosed within leucocytes, it is possible that the bacteria retain their full vigor after being ingested.

It is well known that a suppurating part contains large numbers of leucocytes, and one of the most characteristic events in the inflammatory process is the migration of leucocytes to the point of irritation. This indicates a positive chemotaxis for leucocytes on the part of substances in the inflamed area. Metchnikoff believed that the function of these leucocytes is to destroy the bacteria and to arrest their further progress. On this theory bacteria have often been likened to an invading army and the leucocytes or phagocytes to a force designed to repel their attacks.

It is certain that in some infectious diseases the number of leucocytes, chiefly of the polynuclear neutrophilic variety, in the circulating blood is increased (leucocytosis). This is the case usually in lobar pneumonia and acute suppurative infections. In other infectious diseases there is no leucocytosis; for example, tuberculosis, typhoid fever and malaria. It is interesting to observe that in trichinosis, and more rarely in infection with other animal parasites, the eosino-

---

[1] Metchnikoff, "Comparative Pathology of Inflammation," trans., Starling, 1893.

philic leucocytes become much more numerous in the blood than normally. There is every reason for believing that the leucocytes and other body cells play an important part in combating bacteria. However, the doctrine of phagocytosis, as primarily stated, is insufficient to account for many facts. Metchnikoff himself has modified his original views, to conform with more recent studies.

Ehrlich's Side-Chain Theory of Immunity.[1]—This theory was first proposed to explain the resistance of the body to soluble bacterial poisons, as those of diphtheria and tetanus. In these cases what is known as antitoxic immunity is produced. With the infections in which the poisons occur in the bodies of the bacteria the protective mechanism is different, giving rise to the term bacteriolytic immunity.

Antitoxic Immunity.—Ehrlich first endeavored to explain from a chemical standpoint the action of toxins on cellular protoplasm and the formation of antitoxins. To begin with, the molecules of the protoplasm are to be regarded as being endowed with chemical groups, present in the form of lateral appendages to the molecule, called side-chains. They can be illustrated by the analogies presented by the graphically written formulæ of some complex molecules. It is necessary to conceive of molecules made of an immense number of atoms, and bristling with projecting side-chains. The function of the side-chains is to become attached to other organic molecules with which they have affinities. In this manner they aid in absorbing the substances essential for the nutrition of the protoplasm of cells.

---

[1] The literature of this subject is very extensive. An exhaustive review is that by L. Aschoff, "Ehrlich's Seitenkettentheorie," *Zeitschrift f. allgemeine Physiologie,* 1902.

The following are also of a general character: H. C. Ernst, "Modern Theories of Bacterial Immunity," 1903; Prudden, *Medical Record,* February 14, 1903; Ritchie. *Journal of Hygiene,* Vol. II., 1902; Bergey, *American Medicine,* October 11, 1902.

The side-chains are therefore preferably called *"receptors."*
The numerous receptors which a molecule has are of many
kinds, with affinities for other molecules of different kinds.
Each kind of receptor will then have an affinity for a molecule of a particular kind, which it may be said to "fit," as a
key fits in a lock, although this expression must not be taken
in a literal sense. A receptor to which tetanus toxin might
become attached would not "fit" diphtheria toxin. In order
that toxins may be able to combine with the receptors their
structure must be nearly like that of the food molecules
which the receptors are adapted to receive.

Fig. 48.

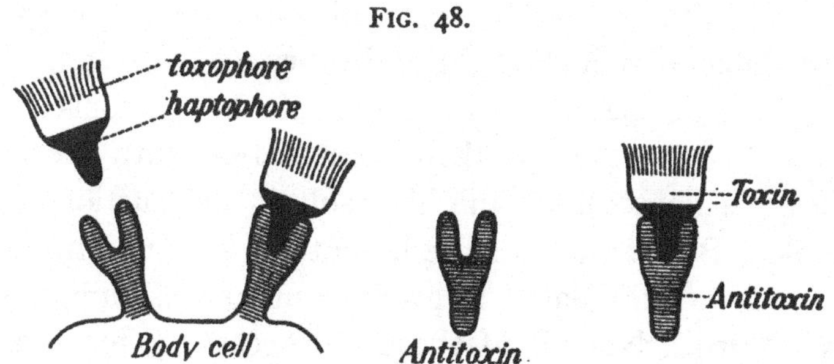

Diagram to Illustrate Side-chain Theory (modified from Aschoff).

Secondly, soluble toxins are to be looked upon as definite
chemical bodies excreted by bacteria, and containing two
essential groups. One group is the *haptophore,* by means of
which the toxin may be linked with the receptors of the
molecules of the cell. The other group is the *toxophore,*
which is capable of destroying the protoplasmic molecule,
after being attached to the receptor of the latter by the
haptophore.

These relations may be represented schematically though
in a very crude manner. In Fig. 48 a portion of a cell is
shown, with receptors. Molecules of toxin, with hapto-
phores and toxophores are near by or attached. At the

right we see a free receptor (antitoxin) and one which has united with toxin to neutralize it.

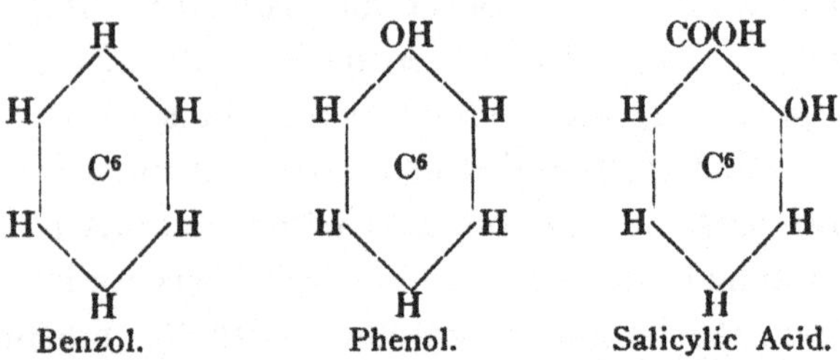

The chemical nature of this conception may be illustrated by recalling the formula of a molecule of benzol. Here one or more H atoms may be replaced with equivalent groups making phenol, salicylic acid and other derivatives.

As the side-chains or receptors of the protoplasm are essential to its existence, their combination with the toxin, through its haptophore, might result in destruction of the molecule. But if the damage be not too serious, the protoplasm may be stimulated to produce numerous similar side-chain groups. Not all of these are necessary for the performance of its functions, and the superfluous ones are thrown off into the surrounding serum. It is well known that many cells of the body exhibit analogous heightened activities under stimulating influences. Such free side-chains or receptors may combine with the haptophorous groups of the toxin, when it will no longer be able to combine with the protoplasm. Thus they act as a kind of buffer in protecting the protoplasm from the attacks of the toxins. They constitute the antitoxic part of the serum.

Numerous experiments have been made which illustrate the probable chemical nature of antitoxic action. A fatal dose of diphtheria or tetanus toxin may be neutralized outside of the body by mixing it with its appropriate antitoxin. Injection of the mixture shows it to be innocuous to animals.

The manner in which toxins combine with protoplasm has

been shown in the case of tetanus toxin. The filtrate from cultures of tetanus bacilli will kill guinea-pigs, presumably by damage to the central nervous system. The same filtrate rubbed up with brain or spinal cord has been found to have lost its toxic proporties. It may be assumed that the poison has combined with the protoplasm of the cells.

Bacteriolysis.—Although the hypothesis of Ehrlich just stated may seem very complicated, more recent studies along similar lines have led to conceptions so intricate, that they can be followed only with the greatest difficulty. The nomenclature of this subject has also become extremely involved, as different words have often been coined to convey a similar idea. In other cases a single word has been used in different senses.

The problems we have still to consider relate to the bacteria which do not produce soluble toxins. In infections with these organisms, attempts to make protective or curative substances analogous to antitoxins have met with little success. The outlook for better results in the future is promising.

The agencies which lead to the destruction of bacteria during the progress of an infectious disease have been studied chiefly in experimental infections in the lower animals. There are numerous bacteria, such as the spirillum of cholera, bacillus of typhoid and bacillus coli communis, injection of which may produce a fatal septicemia in animals. It also usually is possible to render animals immune to them by injections of the dead bacteria, or small doses of living bacteria, or the two in succession. When an animal has been immunized in this manner, injection of living cultures of the same kind is followed by destruction of the bacteria injected. This phenomenon appears to depend on the combined action of two substances. Neither of these is effective without the other. The actual destruction of bacteria (bacteriolysis) is performed by the *complement,* an unstable body, destroyed at quite low temperatures (about 60° C.). The union of the complement with the bacterial cell requires the presence of an *intermediary body,* which is more stable, and more resistant to heat.

Some of the synonyms for these terms are:

Complement,	Intermediary Body,
Addiment,	Amboceptor,
Cytase,	Immune Body,
Alexin,	*Substance Sensibilitrice.*

The nature of complements and intermediary bodies has in a large measure been worked out by the study of immunity to substances not directly related with bacteriology. Among these the blood[1] of various animals and snake venom[2] are notable examples.

The substances developed or increased by immunization are the intermediary bodies, which have, therefore, also been called immune bodies. Their ultimate source must have been in the cells of the host. They may be supposed to have been receptors originally, which were cast off, after the manner of antitoxins. They are specific for the particular kind of organism used in developing immunity. Whether or not various kinds of complements exist is still in doubt.

Metchnikoff holds that while intermediary bodies may occur in blood-serum, the complement exists within leucocytes and other forms of cells. Union of intermediary bodies with the molecules of a bacterial cell, permits destruction of the bacteria by the complement in the phagocyte.

---

[1] Prudden, *Medical Record.* February 14, 1903.

[2] Flexner and Noguchi, " Snake Venom in Relation to Hemolysis, etc.," *Journal Experimental Medicine,* Vol. VI., 1902, p. 277. *University of Pennsylvania Medical Bulletin,* November, 1902.

Disintegration of leucocytes may give rise to the presence of complement in the serum.

An example of the destruction of bacteria in immunized animals is seen in an experiment performed by Pfeiffer. A guinea-pig is given repeated injections of the spirilla of cholera, which have been killed by chloroform or heat. When living spirilla are introduced into the peritoneal cavity of such an immunized guinea-pig they rapidly undergo disintegration. Since no such disintegration takes place when other bacteria than the spirilla of cholera are injected into the animal made immune from this organism, it has been suggested by Pfeiffer that this reaction could be made use of in the diagnosis of that disease. If the serum of the immune animal be introduced along with the cholera spirilla into the peritoneal cavity of an animal not immune, the same disintegration takes place. Furthermore the serum of the immune animal may be heated to 70° C. and will still cause disintegration of the organisms in the peritoneum of a non-immune guinea-pig. The heated serum alone is found inactive. The explanation of this phenomenon appears to be that the serum of the normal animal contains the complement only. The serum of the immunized animal has developed in it the intermediary or immune body, besides the normal complement. The immune body resists heat, while the complement is destroyed by heat. The previously heated serum from the immunized animal, mixed with organisms, finds the necessary complement in the normal animal. The organisms then become disintegrated.

Such disintegration of bacteria is called *bacteriolysis*. The substances which effect it are called *lysins*. It is probable that the development of lysins is one of the most important factors in checking infections and in promoting recovery. The preparation of specific bacteriolytic blood-serum from immunized animals has been attempted for the treatment of various diseases. The partial or complete failure of such efforts may be ascribed to the fact that the serum prepared does not contain both the complement and the intermediary or immune body in proper kinds and proportions to be effective. There is likely to be a deficiency of complement. Investigators are at present very hopeful of being able to make up for this deficiency. The outlook is encouraging at the present time particularly with regard to typhoid fever and dysentery.

Welch's Hypothesis.—When bacteria do not form soluble toxins, we are obliged to suppose that large numbers of them must disintegrate in the body of the host. Thus their intracellular toxins may be liberated and produce the symptoms of intoxication manifested in infections of this character. Welch has also suggested that production of toxins may occur in infection of the living body even when test-tube cultures of the same bacteria contain no soluble toxins or only small amounts. Bacteria growing as parasites in the living body may adapt themselves to the conditions found there. The existence of the bacteria requires them to have some means of combating the substances elaborated by the host for his protection. Bacteria may also be able to form similar substances for their own protection. Such substances, from the standpoint of the host, would constitute toxins.[1]

Nuttall made the important discovery that the serum of the blood deprived of all cells possesses the power of destroying pathogenic bacteria.

The ingredients of the blood-serum that exert the bactericidal influence have been named alexins,[2] they have also been called defensive proteids. The nature of these substances has not been determined with certainty. Vaughan

---

[1] Welch, *Bulletin Johns Hopkins Hospital*, December, 1902.

[2] This word has unfortunately been used partly in another sense, see page 188.

has shown that they may be in part nucleins. He has found that blood-serum contains nuclein and that nuclein has germicidal power. Such substances apparently serve as safeguards to the body against all kinds of bacterial invasion. They have not necessarily a specific action as regards any particular kind of infection.

It now appears that the bactericidal power of blood-serum also depends in a large measure upon the joint operation of complements and intermediary bodies.

The amount of intermediary body appears to be more constant than that of the complement. Flexner[1] found that the blood-serum from cases of chronic diseases had diminished bactericidal power for the staphylococcus pyogenes aureus. His observations were regarded as explaining the liability of these patients to terminal infections. (See page 169.) Longcope[2] has since shown that in such affections as chronic nephritis, cirrhosis of the liver and diabetes the complement, active in destroying bacteria, is diminished. Abbott and Bergey[3] proved that the complement active in disintegrating the blood-corpuscles of another species is reduced in quantity by alcoholic poisoning. Their conclusions are important taken in connection with Abbott's previous work on the effect of alcohol in lowering the resisting power to infection (see page 166).

Agglutinins.[4]—It has been shown that the blood-serum of patients having typhoid fever contains a substance which, when mixed with living typhoid bacilli, causes them to gather into groups or clumps and at the same time to lose their motility. In the great majority of cases no such clumping occurs when the blood of typhoid cases is mixed

---

[1] *Journal of Experimental Medicine*, Vol. I., 1896, p. 21.

[2] *University of Pennsylvania Medical Bulletin*, November, 1902, p. 331.

[3] *University of Pennsylvania Medical Bulletin*, Aug.–Sept., 1902, p. 186.

[4] Sailer, *University of Pennsylvania, Medical Bulletin*, Aug.–Sept., 1902.

with other bacteria than the bacilli of typhoid fever. The nature of this agglutinating substance, as it is called, is not known nor is its significance understood. It has been applied to the diagnosis of typhoid fever, where it is called the " serum-reaction," and will be discussed in connection with the bacilli of typhoid fever. Similar agglutinating bodies form in many other infections. Bacteria that are not motile may nevertheless be agglutinated. Among the infections where such a reaction occurs, the following are noteworthy—with the cholera spirillum, bacillus pyocyaneus, bacillus proteus, bacillus coli communis, micrococcus melitensis, bacillus of glanders, bacillus tuberculosis, diplococcus of pneumonia, bacillus of bubonic plague, and the bacillus of dysentery (Shiga). The protozoön, Trypanosoma (see appendix), is said to become agglutinated in the blood of infected rats. Not all of these have been studied sufficiently to be available for diagnostic purposes. It has not yet been shown that agglutinins are concerned in producing immunity or recovery from infectious diseases. These substances are relatively resistant to heat, not being destroyed by temperatures below 70° C.

Under conditions similar to those under which agglutination occurs, bacteria have been observed to form in long filaments. This is the so-called " thread-reaction." Its significance is uncertain.

Precipitins. — After repeated injections of albumins foreign to it, an animal's fluids undergo still another form of adaptation. Substances are now developed in the bloodserum, which cause precipitates to form where it is mixed with the foreign albumin.[1]   Thus a rabbit may be immu-

---

[1] These reactions have been most fully studied in connection with the precipitation of blood-serum of a particular animal by the serum of another species which has been immunized to the serum of the first animal. The principle is of some importance in medico-legal work for the identification of human blood, as in stains on clothing. See Nuttall, *Journal of Hygiene,* Vol. I., 1901.

nized or adapted to hen's egg-albumen. The rabbit's serum will then precipitate hen's egg-albumen, but no other form of albumin. It may, however, imperfectly precipitate albumen from the egg of a species closely allied to the hen. It has not been shown that this property plays any important part in bacteriology, though that is not improbable.

CHAPTER VIII.

DISINFECTANTS AND ANTISEPTICS.[1]

A **disinfectant** or **germicide** is a substance capable of killing bacteria. The latter term is of more recent development than the former, and apparently needed on account of the loose application of the term disinfectant.

An **antiseptic** is a substance capable of preventing the growth and reproduction of bacteria. It differs from a disinfectant or germicide in that it simply prevents development without actually killing.

A **deodorizer** is a substance capable of so changing a noxious odor that it is less unpleasant to the sense of smell. At the present time the term is usually and properly restricted to those substances which, without disinfectant action, simply replace or destroy an odor.

TESTING ANTISEPTICS AND DISINFECTANTS.

The determination of the antiseptic value of a material is a comparatively simple matter. A virulent culture of the organism used as a test is inoculated into sterile bouillon containing a known quantity of the antiseptic. The process is repeated with varying strengths of the material until the smallest quantity of it capable of preventing growth is determined. This dilution may be considered the antiseptic value of the material in question for the organism used, under the conditions of the test.

[1] By Thomas B. Carpenter, M.D., Assistant City Bacteriologist, Buffalo, N. Y.

Determination of the disinfectant power of a substance is a problem of much greater magnitude, and the method used must be altered more or less to suit the properties of the substance tested. It is obvious that some of the substance tested remains in contact with the organisms in the method given for determining the antiseptic value, and that we do not know whether the bacteria are alive and merely inhibited in growth, or actually killed.

Sternberg's Method.—To a measured quantity of a virulent bouillon-culture of the test-organism is added a known quantity of the substance to be tested. After varying lengths of time inoculations are made from this mixture into culture-media, preferably bouillon, and growth watched for under suitable conditions as to temperature and the like. The shortest exposure of the test-organism to the smallest quantity of the substance is taken as the germicidal value of that substance for the particular organism used.

Koch's Method.—Usually employed for spore-bearing bacteria like the bacillus of anthrax. The hay bacillus is convenient to use when experiments are being made by large classes of students. Small pieces of sterile silk or cotton thread are soaked for some hours in a bouillon-culture of the test-organisms. They are removed, partially dried, and then placed in a solution of known strength of the substance being tested, and exposed for a definite length of time. The thread is removed from the solution, washed carefully in sterile water, planted in bouillon, and growth is watched for. As in other methods, the greatest dilution of the germicide that will kill the test-organism in the shortest time is taken as the germicidal value of that substance for the organism used.

Hill's Method.—The test organism is dried upon the end of a sterile glass rod contained in a sterile test-tube, the end of the rod projecting through a cotton plug. The end of

17

the rod is inoculated with a small amount of a fluid culture of the test organism and allowed to dry. It is then ready to test by exposure to any disinfectant, either liquid or gaseous.

All of these methods are open to serious sources of error, particularly in the testing of powerful germicides. In Sternberg's method, small quantities of the substances tested may be carried over with the organisms, and, if a powerful germicide, in sufficient amount to prevent growth, and thus give erroneous results. In Koch's or Hill's method this factor may be partially obviated by washing in sterile water after exposure to the germicide. This does not remove another source of error, namely, the chemical action that may take place between the substance and the protoplasmic contents of the bacterial cell. This action may extend deeply enough to restrain the growth of an organism for a very long time without actually killing it. When placed under suitable conditions, such union may be broken up and the organism regain its power to develop. It has been suggested that, to remove errors in the above methods, test-cultures containing bacteria supposed to be killed by the smallest quantity of germicide be inoculated into susceptible animals; but Sternberg's experiments in this direction have shown that bacteria may become so altered in virulence by the action of germicides insufficient to kill, that animal inoculation experiments are worthless.

Geppert suggested a valuable modification of these methods while determining the germicidal value of bichloride of mercury. After exposing his test-organism to bichloride of mercury, and before inoculating into bouillon to determine death of the organism, he treated with a dilute ammonium sulphide solution, thus effectually neutralizing any mercury-salt remaining.

Sedgwick developed this method still further, and to him

we are indebted for demonstrating its accuracy and practicability.

Method.—To 15 c.c. of sterile water in a 60 c.c. Erlenmeyer flask add 2 c.c. of a virulent culture of the test-organism. Then add a solution of the substance under investigation in the proportion necessary to give the dilution wished. Mix thoroughly, and allow this " action-flask " to stand as long as it is desired to have the germicide in contact with the test-organism (action-period). Transfer 0.5 c.c. from the action-flask to a flask containing 200 c.c. of a solution of some chemical capable of decomposing the substance being tested with the formation of inert or insoluble compounds. In this " inhibition-flask " the strength of the solution should be such that molecular proportions of the chemical are present in sufficient quantity to combine with all the germicide carried over. The inhibition-flask is shaken for 30 seconds, and 1 c.c. transferred from it to 100 c.c. of sterile water in another, the " dilution-flask." After two minutes, three agar tubes are inoculated with 1 c.c. each from the dilution-flask, plated, and growth watched for.

Control-experiments should be performed to determine that the dilution of the test-culture is not too great when carried through the three flasks. It likewise should be determined that the inhibiting chemical has no effect on the bacteria.

What the inhibiting chemical shall be must be determined for each individual case. For salts of the heavy metals ammonium sulphide answers well; for mercury salts, stannous chloride may be used; for formaldehyde, ammonium hydrate; for carbolic acid, sodium sulphate.

The testing of gaseous disinfectants, such as sulphur dioxide and formaldehyde, must be conducted under conditions as nearly parallel to actual practice as possible. The test-organisms may be exposed on threads or glass rods, and acted upon by a known volume strength of disinfectant for a known length of time. Subsequent treatment of the organisms with a suitable inhibitor is necessary when possible, and should growth occur in the cultures following, the test-organism should be recognized in order that possible contamination by extraneous organisms may be excluded.

In determining the value of germicides for sterilizing ligatures, the students can apply methods based on the foregoing principles. Great care and ingenuity are necessary to arrive at correct conclusions, particularly in the case of animal tendons. In many instances quite stable compounds are formed between tendon and germicide, and living organisms may be so imbedded in such a substance that subsequent growth in a test-culture is impossible. The use of a suitable inhibitor, and, prior to final culture-tests, a prolonged soaking in sterile water, will promote the accuracy of the results.

CHEMICAL DISINFECTION.[1]

Heat properly applied is the simplest and at the same time the surest disinfectant (see Part I., Chapter II.); but for many purposes it cannot be used, and we have recourse to those chemicals that practice and investigation have shown to be of value. *The efficiency of chemical disinfectants as ordinarily used is over-rated.* An immense number of substances possess germicidal properties, but unfortunately, the majority are objectionable in that they are expensive, intensely poisonous, or so corrosive that damage may be done to articles of value with which they may come in contact.

In the following pages only those substances which are in common use or seem to be of special value will be considered.

[1] For fuller details on this subject consult Rosenau, "Disinfection and Disinfectants," 1902.

Mercuric Chloride or Bichloride of Mercury.—This substance is probably more commonly used than any other one disinfectant. In the strength of 1–1000 it will sometimes kill the spores of anthrax within a few minutes (see Bacillus anthracis, Part IV.). It is claimed that its affinity for albuminous bodies, and the readiness with which it combines with such substances, detracts from its value for some purposes. On the other hand, many observers claim that the albuminous combinations formed under such circumstances are soluble in an excess of albuminous fluid, and that its value as a germicide is not affected thereby. To obviate this possible difficulty it is customary in practice to combine the bichloride of mercury with some substance that will prevent the precipitation of the mercury salt by albumin. For this purpose 5 parts of any one of the following substances to 1 part of bichloride of mercury may be used—hydrochloric acid, tartaric acid, sodium chloride, potassium chloride or ammonium chloride. A very practical stock-solution for laboratory purposes has the following composition:

Hydrochloric acid .... 100 c.c.
Bichloride of mercury .... 20 grams.
5 c.c. in a liter of water makes a solution of about 1–1000 strength.

Mercuric Iodide.—An extremely high antiseptic value has been placed on this substance by Miquel, who claims that the most resistant spores are prevented from developing in a culture-medium containing 1–40,000. In combination, as potassio-mercuric iodide, it has been used in soaps (McClintock) with very favorable results. The substance is not extensively employed, and further investigation is necessary to determine its true value.

Attempts are being made to manufacture combinations of mercury and other powerful metallic germicides with organic acid and basic bodies, the purpose being to utilize the metallic base in greater strength without injury to the living tissues. Such compounds are exemplified by *mercurol,* said to be a combination of mercury with nucleinic acid, and to possess active germicidal properties, great penetrating power and no injurious effect on living tissue. It is also said to have a particularly destructive action upon the gonococcus.

Silver Nitrate.—This salt probably occupies the next position to the bichloride of mercury in disinfectant power. Behring claims it to be superior to bichloride of mercury in albuminous fluids. The anthrax bacillus is killed by a solution of 1–20,000 after two hours' exposure. At least forty-eight hours' exposure to a 1–10,000 solution is required to kill the spores of anthrax. It is very irritating, and possesses strong affinities for chlorides, forming with them insoluble chloride of silver, a salt without germicidal value. For these reasons the use of silver nitrate is limited. In the solutions usually employed for douching the cavities of the body, the available silver nitrate is immediately converted into the insoluble chloride, and little if any germicidal action takes place. To this fact may be ascribed the varying clinical results reported.

Many semi-proprietary silver compounds are on the market, introduced to replace the nitrate and its objectionable features. The most important are argentamin, argonin, protargol and argyrol, all organic silver combinations. They do not combine with chlorides, are less irritating than the nitrate, and, not coagulating albumin, they possess greater penetrating power. Clinical reports and investigations have been so contradictory thus far that their value cannot be readily estimated.

Carbolic Acid.—One of the most important and most widely-used disinfectants. It is usually employed in strengths of from 1 to 5 per cent. A 3 per cent. solution will sometimes kill the spores of anthrax after two days' exposure (see Bacillus anthracis, Part IV.). In the absence of spores the anthrax bacillus is destroyed by a 1 per cent. solution in one hour. The less resistant pus cocci are destroyed rapidly by a 2 per cent. solution. Combination with an equal proportion of hydrochloric acid enhances the efficacy of carbolic acid to a marked extent. This is due to the prevention of albuminous combinations, thus allowing greater penetration of the disinfectant.

Many other substances closely related to carbolic acid are used and possess marked germicidal properties. Among them may be mentioned creolin, cresol and lysol. They are all slightly superior to carbolic acid in actual germicidal value.

Aniline Dyes.—Many of these substances possess germicidal properties, notably pyoktanin (methyl-violet). A solution of 1–5000 will kill the anthrax bacillus in two hours. A much stronger solution, 1–150, is required to kill the typhoid bacillus in the same time. Malachite-green is said to possess even greater germicidal value than pyoktanin. Methylene-blue also possesses considerable germicidal power.

Formaldehyde.—A gaseous substance placed on the market in a 40 per cent. aqueous solution. Remarkable claims have been made for this substance, and numerous investigations have shown it to possess, both in the liquid and gaseous forms, wonderful disinfecting power under certain conditions. It is a noticeable fact that the more recent the investigation the lower the value placed upon it. In solutions of 1–1000 an exposure of twenty-four hours is necessary to destroy the staphylococcus pyogenes aureus, while 1–5000 is sufficient to restrain its growth (Slater and Rideal). Its use in a gaseous form as a house-disinfectant is by far the most important application at the present time.

Harrington's investigations have shown that an atmosphere produced by vaporizing 435 c.c. of formalin (40 per cent. aqueous solution of the gas) in 1000 cubic feet of air space, equivalent to 1 quart to a room 15 feet square and 10 feet high, will destroy all exposed organisms in half an hour; when protected by one fold of cotton-cloth, an exposure of one and one-half hours is necessary. In a perfectly dry atmosphere the gas penetrates slightly, and will disinfect through one layer of cotton-cloth; in a moist atmosphere no penetration can be obtained.

In vaporizing the gas many methods have been employed. Simple evaporization of solutions without heat cannot be relied upon, for the solid, polymerized paraformaldehyde is easily formed under these circumstances. Better results can be obtained with the aid of heat, although polymerization is apt to occur unless evaporation is rapid. To produce the best results it has been found necessary to use special forms of lamps or generators for its production, a few of which are mentioned below.

Sanitary Construction Company's Lamp.—This lamp consists of a tank to hold the formaldehyde solution, and a spiral tube by which the solution is slowly conducted through a flame and vaporized. The necessary amount of solution is placed in the tank and the apparatus started, outside the room, the gas being conducted through the keyhole by a suitable tube.

Trillat Autoclave.—A small silver-lined pressure-boiler, fitted with lamp, safety-valve, pressure-gauge, thermometer and escapement-tube. The necessary amount of formaldehyde solution is placed within the apparatus, together with an equal amount of 20 per cent. solution of calcium chloride; the addition of the latter salt is to prevent formation of the solid polymeric modification, the so-called paraform. The autoclave is closed and heated from below to a temperature of 135° C. The escapement-valve is then opened carefully and the gas allowed to enter the room slowly through the escapement-tube, which has meanwhile been passed through the keyhole. About thirty minutes are required to discharge all the gas from 500 c.c. of solution. If the temperature has not been allowed to go above 135° C. the gas will contain but little moisture and possess its maximum efficiency.

Schering Lamp.—This lamp is intended to utilize paraform or para-formaldehyde, a polymeric modification of formaldehye, occurring as a white salt. It is decomposable by heat, yielding formaldehyde gas. It is placed on the market in the form of tablets, each one of which yields a definite amount of gas. The lamp consists of a small iron tray for the reception of tablets, and so arranged above the heating-apparatus that sufficient draught is created to carry off the gas as rapidly as formed. In operating, a sufficient number of tablets are placed on the tray, the lamp lighted and placed in the room to be disinfected.

Methyl-Alcohol Lamps.—Several of these lamps are on the market, all operating on the well-known principle of the oxidation of wood-alcohol to formaldehyde when the alcohol is vaporized by projection against a heated, platinized, asbestos disk. In operating such an apparatus, the alcohol is lighted until the asbestos disk becomes hot. The flame is then extinguished; the heat from the disk is sufficient to vaporize the alcohol, which undergoes oxidation and keeps the disk at a red heat. When the apparatus is operating in a satisfactory manner the room is closed and disinfection allowed to proceed. It must be said, however, that it is difficult to estimate or control the amount of formaldehyde evolved in generators of this type.

Formaldehyde Candles.—Mixtures of para-formaldehyde and paraffin or other combustibles, which may be moulded into candles, each enclosed in a tin case, make a convenient apparatus to generate formaldehyde gas for room disinfec-

18

tion. The candle is placed in a suitable fire-proof dish, it is then ignited, and generation of the gas is allowed to proceed in the tightly closed room.

Sulphur Dioxide.—This substance is used extensively for house disinfection, and is usually prepared by burning sulphur. Much difference of opinion exists regarding the value of it as a disinfectant. The spores of anthrax are not killed by several days' exposure to the liquefied gas. Anthrax and other bacilli are destroyed in thirty minutes when exposed on moist threads in an atmosphere containing one volume per centum of the gas. An exposure of twenty-four hours in an atmosphere containing four volumes per centum of the gas will destroy the organisms of typhoid fever, diphtheria, cholera and tuberculosis. The presence of moisture greatly enhances the activity of the disinfectant, owing to the formation of the more energetic sulphurous acid.

For the destruction of insects, such as mosquitoes, this agent is superior to formaldehyde. Its application for this purpose is important in preventing the spread of yellow fever and malaria.

In practice, at least 3 pounds of sulphur per 1000 cubic feet should be used, and moisture must be present. This latter requirement can be fulfilled by evaporating several quarts of water within the tightly closed room just prior to generating the gas. In using powdered or flowers of sulphur, the necessary amount is placed on a bed of sand or ashes in an iron pot, which should rest on a couple of bricks in a pan or other vessel containing an inch or two of water. The sulphur is ignited by means of some glowing coals, or by moistening with alcohol and applying a a match. Difficulty is often experienced in keeping the sulphur burning, and for this reason it is surer and more convenient to use the so-called sulphur candles now on the market. In operating with these, a sufficient number are placed on bricks in a pan of water and the wicks lighted. Liquefied sulphur dioxide may be used, and can now be obtained in convenient tin receptacles containing a sufficient quantity for the disinfection of an ordinary room. The can is opened by cutting through a soft metal tube projecting from the top. The fluid vaporizes at the room temperature, and it is simply necessary to place the can in a convenient porcelain dish and allow the fluid to evaporate.

Sulphur dioxide is objectionable on account of its lack of power when dry, and on account of its corrosive action on metal and its bleaching effect on hangings and draperies in the presence of moisture; it is, therefore, preferable to use formaldehyde when possible.

Chlorine.—A very active gaseous disinfectant, particularly in the presence of moisture. An atmosphere containing 1 per cent. of the dry gas is fatal to anthrax spores in three hours. The anthrax bacillus is killed in twenty-four hours by exposure to a moist atmosphere containing the gas in the proportion of 1–2500. The bacillus of tuberculosis is killed by an exposure of one hour to a moist atmosphere containing the gas in the proportion of 1–200. Extremely minute quantities in solution will prevent the development of putrefactive organisms. The substance has been used for house and ship disinfection, but is now seldom employed on account of its extremely irritating properties and the difficulty of handling it.

Bromine.—Used in the gaseous and liquid form. The dry vapor possesses but little disinfectant power; when moist it is much more efficient. In saturated aqueous solution it will kill the anthrax bacillus in twenty-four hours.

Calcium Hypochlorite, usually known as *Chloride of Lime.*—This is a most practical and valuable disinfectant, depending for its efficiency on the available chlorine contained in it. Its alkalinity favors penetration, and for many purposes it cannot be excelled. A 1 per cent. solution will destroy anthrax spores in one hour. A solution of the same strength will disinfect typhoid stools in ten minutes.

Lime.—The addition of 0.1 per cent. of unslaked lime to fluid-cultures of the typhoid bacillus and cholera spirillum will render them sterile in four or five hours. Typhoid dejecta are sterilized in six hours by the addition of 3 per cent. of slaked lime; the addition of 6 per cent. will accomplish the same result in two hours. A convenient form for practical use is an aqueous mixture containing 20 per cent. of lime—so-called milk of lime. Typhoid and cholera dejecta are sterilized in one hour after the addition of 20 per cent. of this mixture. In practice it is safer to use a considerable excess of lime. From the foregoing facts it would seem probable that lime or whitewash as ordinarily applied would possess disinfectant properties. Experimental work has demonstrated this to be a fact. The organisms of anthrax, glanders and the pus cocci were destroyed within twenty-four hours by one application. For spore-forming organisms and the bacillus of tuberculosis the power is not so great, the latter organism not being destroyed by three applications of the whitewash. This is due, perhaps, to the large amount of fatty matter in the bacillus of tuberculosis, and suggests the possibility of enhancing the efficacy of the lime by the addition of a small proportion of caustic alkali.

Hydrogen Peroxide.—This substance is placed on the market in solutions varying in strength from 10 to 30 volumes; the mode of expression indicating that corresponding solutions will liberate ten to thirty times their volume of oxygen when appropriately treated. It possesses the property of rapidly oxidizing purulent secretions, and on this account is much used for cleansing infected wounds. It deteriorates in strength so rapidly that only fresh solutions of known strength should be used.

Potassium Permanganate.—Koch asserts that a 3 per cent. solution will destroy anthrax spores in twenty-four hours, but that a 1 per cent. solution cannot be depended upon to kill pathogenic organisms. Its disinfectant value in practice is very low on account of its ready decomposition by inert material. In the dilute solutions usually used for medicinal injections and irrigations no disinfectant action occurs.

Iodoform.—This substance possesses little if any disinfectant power. It is mildly antiseptic in moist wounds, due to the gradual liberation of small quantities of iodine.

Boric Acid.—This material possesses practically no disinfectant power. It is a mild antiseptic when applied as an undiluted powder to wounds. A saturated aqueous solution is much used, and is weakly antiseptic.

Essential Oils.—Many of these bodies possess germicidal value, notably the oils of cinnamon and cloves. The oil of mustard is also a valuable disinfectant, but so irritating that the pure oil cannot be used. The use of powdered mustard in the autopsy-room will remove the foul odor from the hands more rapidly and completely than any other means.

Coal Oil or Petroleum.—While the disinfectant value of this substance is slight, its use in destroying the larvæ of insects, such as the mosquito, has given it an important position in preventing the spread of malaria and yellow fever. A small amount poured on a stagnant pool rapidly spreads over the surface, and effectually destroys such larvæ.

Ferrous Sulphate (Copperas).—This salt has been much used, but possesses only feeble disinfectant powers. A 3 per cent. solution requires three days to kill the bacillus of typhoid fever. On account of its affinity for ammonia and sulphides it is an efficient deodorizer for temporary use, but cannot be relied upon to kill the bacteria producing the noxious gases.

Cupric Sulphate (*Blue Vitriol*).—This salt is quite an efficient disinfectant. In a solution of 1–3000 the spirillum of cholera is destroyed in ten minutes. A 5 per cent. solution will kill the typhoid bacillus in ten minutes. A solution of from 2 to 3 per cent. in strength can be relied upon to destroy all pathogenic organisms that do not form spores.

Zinc Sulphate.—This salt is a very feeble disinfectant. Pus cocci are not destroyed in two hours by a 20 per cent. solution. As a deodorizer it has about the same value and acts in the same way as ferrous sulphate.

Zinc Chloride.—A 2 per cent. solution will kill pus cocci after an exposure of two hours. It is therefore a much more powerful disinfectant than the sulphate.

Disinfection of Dejecta and Urine.—A 4 per cent. solution of calcic hypochlorite (chloride of lime) is most efficient and rapid for this purpose. A convenient solution contains 6 ounces of the salt to 1 gallon of water. The excreta should be received in a suitable vessel and immediately mixed with an equal bulk of the disinfectant. The contents of the vessel should be allowed to stand for one hour before emptying. A 20 per cent. milk of lime is just as efficient, and possesses the advantage of cleanliness and lack of odor. It should be used in the same quantity and allowed to act for the same length of time. A 5 per cent. solution of carbolic acid may be used, but should be allowed to act for at least four hours.

Disinfection of Sputum.—The chemical disinfection of tuberculous sputum is somewhat difficult on account of the large amount of albumin in it and the fatty matter associated with the bacillus of tuberculosis. Dilute solutions of bichloride of mercury are apt to be decomposed and rendered inert by the albumin. Carbolic acid is open to the same objection, but its combination with hydrochloric acid can be used successfully in a strength of 5 per cent. each. Milk of lime cannot be relied upon for this purpose. A 4 per cent. solution of calcic hypochlorite (chloride of lime) is the best for general use, and the spit-cup should be kept nearly full of this solution. Sputum may also be disinfected by exposure to the action of steam in the steam sterilizer or by boiling for 15 minutes. If napkins or old pieces of cloth are used for the reception of sputum they may be immediately destroyed in a fire.

Disinfection after Postmortems.—After autopsies on infectious cases it is necessary to disinfect the table and fluid products coming from it prior to emptying into the sewer. The table may be successfully disinfected by a liberal sprinkling with 4 per cent. calcic hypochlorite solution. All fluids should be treated with an equal quantity of the same solution. The table should not be cleaned for at least one hour after application of the disinfectant. The same rule applies to the disinfection of the fluids—an exposure of at least one hour to the disinfectant before final disposition.

The Cadaver in Contagious Diseases.—In cases of death from a contagious disease all the orifices of the body should be packed with cotton soaked in a strong solution (1 to 500) of bichloride of mercury, the skin washed with a 1 to 1000 solution, and the cadaver wrapped in a sheet wet with the same. The funeral should be private and the body disposed of within twenty-four hours, preferably by cremation.

House Disinfection.—After infectious disease it is essential that the house or the apartment in which the patient has been confined should be disinfected. It is rarely necessary to carry out the process in more than two rooms; but should it be so, the process can be applied to the whole house.

After thorough bathing of the patient, preferably with an antiseptic soap, the individual should be wrapped in a clean sheet and removed to a clean room. All articles or materials that are of little value should be destroyed. All bedding, towels and the like should be placed in wooden tubs and covered with a 1–1000 solution of bichloride of mercury. The room should then be made as nearly air-tight as possible; this can be accomplished by pasting strips of paper over registers, cracks, spaces between window-sashes and the like. Formaldehyde gas is then passed through the keyhole into the room (or it may be generated by formaldehyde candles) in sufficient quantity to destroy the infectious element. The room should be sealed for at least twelve hours, after which time it may be opened and aired. The process is completed by washing all exposed surfaces in the room with 1–1000 bichloride of mercury. This latter requirement is not essential if the gaseous disinfection has been complete, but since we have no absolute knowledge on this point, the secondary washing should be carried out. This method can be considered reliable for surface disinfection, but for the interior of mattresses and stuffed furniture-cushions it is not certain. In all cases where absolute disinfection is demanded, such articles must be ripped apart and loosely exposed to the gas. They may be disposed of by fire or sterilized by steam under pressure. The latter method must necessarily be a matter of municipal control, and can only be carried out by means of suitable apparatus in the hands of a municipal disinfecting corps. Instead of formaldehyde, sulphur dioxide may be used for room disinfection, but in the light of present knowledge the formaldehyde method is superior.

CHAPTER IX.

THE PREPARATION OF INSTRUMENTS, LIGATURES, DRESS-INGS, ETC., FOR SURGICAL PURPOSES.[1]

THE purpose of this chapter is to explain the application of the principles set forth on the preceding pages to surgical technique. It has been shown that all objects about us may have bacteria on them, and that bacteria are present on all the surfaces of our bodies that come in contact with the air. All the care that is needed in working with bacteria in the laboratory, and more, must be exercised in surgical operations. Everything that has not been sterilized must be regarded as having the possibilities of infection in it. After the hands have been cleansed, if they touch the clothing or furniture, they must be cleansed again. If a sterilized instrument falls on the floor, it must be sterilized again. The same applies to dressings, sponges, ligatures, or anything which is to be used about the wound.

The value of chemical germicides has probably been overrated in the past. They are used only to destroy the bacteria on living tissues and on articles that would be damaged by heat. They give less reliable results than boiling. Wherever boiling or steam sterilization is permissible, it should be used. With materials that may contain a small quantity of substance in which bacteria can grow, the fractional method of sterilization should be used (see page 63). With glass and metallic objects, obviously a single boiling can accomplish as much as boiling on three consecutive days.

[1] By Marshall Clinton, M.D., Instructor in Clinical Surgery, Medical Department, University of Buffalo.

19

The failures in the practice of aseptic surgery are generally due to the hands of the operator and assistants and the skin of the patient.

The following formulæ have been selected from the many published as they are successfully used by many surgeons, and meet the theoretical grounds of bacteriology as far as is possible with our present knowledge.

Sterilization of Hands.—There is no known method for perfect sterilization of the human skin. A close approach to sterility is reached by any one of the methods that has as its basis mechanical cleanliness.

Park's method: (1) Hands and forearms are thoroughly rubbed with a mixture of green soap and cornmeal, which serves to remove all the loose dirt and epithelium. Rinse carefully until hands and forearms are clean. (2) A paste of mustard flour and cold water is rubbed into the hands and forearms until they begin to sting. (3) Rinse in running sterile water; then soak in a hot 1–1000 bichloride of mercury solution for a few minutes, the fluid being well rubbed into the skin.

Fürbringer's method: (1) Thorough scrubbing of the hand and forearms with soft soap, water and a nail-brush for at least three minutes, especial attention being paid to the nails. (2) Removal of all fat and debris by rubbing hands and forearms while immersed in 95 per cent. alcohol. (3) Rinsing of hands and forearms in a 1–1000 bichloride of mercury solution, rubbing the fluid well into the skin.

Schatz's method: (1) Hands and forearms are cleansed by brisk scrubbing with soft soap and a clean brush for from three to five minutes. (2) Soaking in saturated solution of permanganate of potassium at a temperature of 110° F. until the hands and forearms are a deep mahogany brown. (3) Immersion in a saturated solution of oxalic acid, temperature of 110° F. until the skin is entirely decolorized.

(4) Rinsing with sterile lime water to rid of excess of acid. (5) Washing in 1–1000 bichloride of mercury solution for one minute.

Weir's method: (1) Hands and forearms are scrubbed as in other methods. (2) A scant tablespoonful of chlorinated lime is moistened with enough warm water to make a thick paste. This is carefully rubbed into hands and forearms. (3) A piece of carbonate of soda one inch square and one-half inch thick is crushed and rubbed into the paste. From three to five minutes are thus employed. (4) Rinsing in sterile water and washing in a solution of $\frac{1}{2}$ of 1 per cent. of ammonia, removes the odor of chlorine.

E. R. McGuire[1] states that prolonged scrubbing with frequent changes of brushes in running sterile water will give the nearest approach to sterility. The use of antiseptics on the hands is not to be relied upon, for their precipitation by chemicals or normal tissue fluids may break up their combination with bacteria that were considered inactive or dead, but are not so in reality. He suggests the use of hot-air, by cabinet-bath or Kelly hot-air apparatus, to "sweat" out of the glands in the skin as much as possible of their contents before the skin is cleansed.

Prolonged soaking of the skin in a soap poultice or strong antiseptic may damage and irritate the tissues, so that it is not advisable to prepare the field of operation more than twelve or twenty-four hours before the time set for an operation.

Maylard[2] recommends the sterilization of the skin by inunctions of oleate of mercury. The method employed is as follows: (1) Cleanse the skin in the usual way with soap and water. (2) Anoint freely and widely with hydrated lanolin-oleate of mercury, 20 per cent., and rub in; smear a piece of gauze with the same and leave until a second inunction is performed twelve hours later. Every case

[1] *American Medicine*, February 28, 1903.
[2] *Annals of Surgery*, January, 1902.

should be treated for at least twenty-four hours before operation ; preferably forty-eight hours should be given, with at
least two separate periods of " rubbing in " for about ten
minutes on each occasion.   (3) On the operating table the
piece of gauze is removed, and the superfluous ointment
rubbed off with a piece of sterile gauze.

To Prepare the Field of Operation.—Wash with green
soap and water, scrubbing thoroughly and carefully, paying
particular attention not to scrub hard enough to render the
skin tender or to make abrasions.   Shave parts with clean
razor.   Wash with ether and alcohol, to remove debris and
epithelium, and cover with a sterile towel.   If the skin of the
patient is thick a soap poultice may be left on, care being
taken to see that the skin does not become macerated.   After
the patient is anesthetized the field is briskly scrubbed with
sterile brushes, soap, and water, washed with 1–1000
bichloride of mercury solution and covered with sterile
towels.

It is important to remember that during an operation
patient, operator and assistants, may perspire and that in this
way fresh masses of bacteria from the deeper parts of the
glands may be brought to the surface of the skin.   Careful
attention must be paid to maintaining cleanliness during an
operation.   The patient's skin is kept covered with sterile
towels, changed as often as they become soiled.   For the
surgeon's and assistant's hands *rubber gloves* do this perfectly.   If an operator or assistant finds that the hands perspire during an operation the use of rubber gloves becomes
essential.   Rubber gloves may be sterilized by boiling.

Instruments are best sterilized by contact with super-
heated steam, or steam under pressure for ten minutes, or
by boiling in a 1 per cent. carbonate of soda solution.   If
soda is unavailable use water that is actively boiling, as this
avoids spotting and rusting of the instruments.   Imme-

diately after use instruments should be thoroughly scrubbed with a brush and washed with soap and hot water and boiled, before being replaced in the instrument case.

The practice of passing an instrument through a flame a few times cannot be relied on to destroy the bacteria that may be present.

Aspirating syringes, needles, trocars, drainage tubes and glass nozzles are best sterilized by boiling for ten minutes. If syringes have leather washers (which should be avoided) they may be cleansed with hot water and soap, rinsed with alcohol, filled and refilled with boiling water ten or more successive times, and placed in a 1–40 carbolic acid solution.

Instrument trays, ligature dishes, basins for sponges, etc., are to be sterilized by boiling for ten minutes, and protected from dust with sterile towels.

Catgut is made from the intestines of sheep, and sheep are subject to anthrax infection, while tetanus bacilli may occur in the intestine. Therefore, catgut must be sterilized by some method that will kill the spores of these organisms if present (see Bacilli of anthrax and tetanus, Part IV.). There are many methods devised for the preparation of sterile catgut that have as a basis an incorporation within the catgut of some antiseptic. They are open to the objection that any antiseptic introduced into the tissues acts as an irritant, aside from the fact that organisms may be liberated from partially absorbed catgut. This is seen in cases of late suppuration—ten to fifteen days after an operation.

Catgut comes in sizes from double zero up to No. 8. The sizes mostly used are 0 to 4. Catgut when ready for use should be smooth, soft, pliable, and very strong; wiry catgut is apt to cut through tissues.

Cumol method.[1] The catgut is rolled on glass spools, and these put in a glass beaker. The bottom of the beaker is

[1] Clark and Miller, *Bulletin Johns Hopkins Hospital*, Vol. XI., September, 1900.

covered with a layer of cotton on which the catgut rests. The beaker is heated with a Bunsen burner over a sand-bath. The top of the beaker is covered with a piece of cardboard. Through a hole in the center of the cardboard a thermometer passes. Heat is now applied to the sand-bath, and the temperature of the catgut slowly raised to 80° C. In this manner all moisture is driven out of the catgut. This degree of heat is maintained for one hour. Cumol[1] at a temperature of 100° C. is now added to the beaker, completely covering the catgut. The beaker should be covered with copper-wire netting to prevent ignition of the cumol, which is very inflammable. The temperature is then increased to 165° C., and kept at this point for one hour. The fluid is now poured off, and the catgut allowed to dry in the beaker on the sand-bath at a temperature of 100° C. for two hours. It is then transferred to sterile jars or test-tubes until needed, or it may be preserved in sterile alcohol.

Formaldehyde catgut.[2] Three-quarter-inch glass spools are notched on each flange. The catgut is wound upon the spool tightly in one layer, and evenly, the ends passing over the flange of the spool in the notch; the longer end, after passing through the notch, goes through the barrel of the spool and is securely tied to the shorter end which has passed over the other notched flange. By thus winding the gut there will be enough for one or two ligatures or sutures of good length. Gut prepared by this process tends to contract forcibly, and on account of this strain must be held securely or it will shrink and be useless. The object of winding in a single layer, evenly, is to prevent over-lapping or crossing of one strand over another. If in

---

[1] Cumol is a fluid hydrocarbon, with a boiling point somewhat above 165° C. It dissolves the fat in catgut. After boiling it has a brown color.

[2] Frederick, *American Journal Obstetrics,* Vol. 89, 1899.

the process of soaking in formaldehyde and the consequent shrinking, one strand crosses another, the one next to the glass will be so pressed upon as to prevent hardening at that point. When the gut is boiled later in water, that point will gelatinize and break at the least strain. Formaldehyde comes in a 40 per cent. solution. We use a 3 per cent. solution, pouring one part of the 40 per cent. formaldehyde solution and thirteen parts of water into a wide-mouthed bottle. Immerse the spools in this solution for periods of time varying with the size of the gut. No. 0 is left in one hour. Nos. 1, 2 and 3 are given three, five, and seven hours respectively. If left too long in the solution the gut will become too hard, too brittle, and the strength will be impaired. Wash in running water for a longer time than it was in the formaldehyde solution. Up to this time the gut has not been sterilized. It has undergone a chemical change whereby it may be boiled without spoiling it. The sterilization of the gut consists in boiling for fifteen minutes, with the receptacles in which it is to be kept. With sterile forceps place the spools, each size by itself, in wide-mouthed-ground-glass-stoppered bottles or in rubber-sealing fruit jars, sterilized by boiling. Pour over the gut clean 95 per cent. alcohol with 8 to 10 per cent. of glycerine. To sterilize the glycerine it should be placed in a bottle in water and raised to the temperature of boiling water for half an hour.

To make *chromicized catgut* wind the spools as before. Place the spools in a solution of: bichromate of potassium, 1.5 grammes; glycerine and carbolic acid each 10 c.c.; water 1 liter. Allow them to remain in this solution for twenty-four hours. Take out and drain, allowing them to dry for a few hours. Then place in the formaldehyde solution and put through the same process as with formaldehyde catgut.

Kangaroo tendon, owing to its slow absorption, is used as

a heavy retaining suture, and is prepared by washing the strands in ether to free from fat. Soak in a 4 per cent. solution of chromic acid for twenty-four hours. Then sterilize by the cumol method.

Silk may be sterilized by the fractional method (see p. 63) as this does not impair the strength as does boiling.

Silkworm gut is prepared by steam sterilization by the fractional method or by boiling in plain water for one half hour. It should not be boiled in soda solution as this spoils the gut.

Horsehair strands are cut into two-foot lengths, washed with soap and water and sterilized with steam by the fractional method. They make a very fine suture and are used where an inconspicuous scar is particularly desirable, as on the face. Only the finer grades are used for this purpose.

Silver wire.[1] This material has the advantage over other suture materials of having a germicidal or at least a restraining influence on bacteria. If we remember that absolute sterilization of the skin is not possible by any means, we must see that in silver wire as a skin suture we have a safe and valuable material. Recent annealing by heating to a dull red increases the flexibility of the wire but almost totally destroys its germicidal property. This will reappear in a month and is not disturbed by boiling. Therefore prepare it by boiling for ten minutes in the 1 per cent. soda solution.

Sponges. The best absorbents to use in surgical work are those whose sterility is undoubted. Pads of gauze are easily sterilized by steam as for dressings. Sea sponges[2] may be prepared by beating with a wooden mallet to remove sand and dirt. Soak in a 1–64 solution of hydrochloric acid for

[1] Bolton, *Transactions Association American Physicians,* 1894.

[2] McBurney, "International Text-Book of Surgery," June, 1900, p. 284.

twelve hours to remove lime deposits. Wash in running warm water. Soak for fifteen minutes in a saturated solution of permanganate of potassium, then place in a saturated solution of oxalic acid until they are perfectly bleached. After immersion for half an hour in this solution rinse thoroughly in sterile water and put in a 1–1000 bichloride of mercury solution for twenty-four hours. Remove and place in 1–20 carbolic acid solution until required for use. At operation remove from solution, rinse out in normal salt solution, and place in receptacle filled with salt solution. If sea sponges are used on a septic case they should be thrown away and no attempt made to resterilize them. If used on clean cases they may be resterilized as above.

Dressings. The two materials universally used to dress wounds are " gauze " or cheese-cloth and absorbent cotton. If they are properly sterilized, the impregnation of gauze or cotton with antiseptics does not add to their value. Gauze is usually cut in portions one yard square and folded in pieces called compresses. A number of compresses are wrapped with a piece of cotton cloth and the edges stitched loosely into a closed bundle. After sterilization by the fractional method, the bundles can be placed in sterile jars or receptacles and each bundle removed as needed. Ripping open the stitches gives untouched sterile bundles of compresses, convenient and handy for using. Cotton is sterilized by the fractional method in rolls or bundles as for gauze. These dressings should be warmed before being placed in the steam sterilizer or they will be unnecessarily wet when removed.

Irrigating Solutions. Chemical germicides, such as bichloride of mercury when in solution, cause necrosis of tissue. Plain sterile water causes maceration of epithelium. Normal salt solution is the least irritating to the tissues and is the one most generally employed for irrigating purposes. It is .6 per cent. sodium chloride, prepared roughly by adding a teaspoonful of salt to the pint of water. This solution may be sterilized by boiling for half an hour on three consecutive days. It does not injure tissue, and may be freely used in operations for irrigating. It has no germicidal or antiseptic properties.

Accident wounds are generally lacerated or contused and may contain pathogenic bacteria. They should be promptly and carefully cleansed with sterile salt solution, wiped with sterile gauze and if necessary scrubbed vigorously with sterile soap and brush to remove all infectious dirt. When there is any doubt of this being accomplished it is better to dress such wounds wide open, filled with sterile gauze, for forty-eight hours or more. Retained blood clots form a good medium for the development of bacteria so that drainage for a day or two is safer in doubtful cases.

Infected wounds. There is no known method for promptly sterilizing infected wounds without destroying tissue. An infected wound, if the infection be not too deep, may be sterilized by cauterizing with pure carbolic acid.

Care must be exercised in the application of antiseptic solutions in infected wounds for the antiseptic rarely penetrates as deeply into the tissues as the bacteria are found, therefore, further necrosis of tissue and mechanical cleanliness are about all they accomplish.

After-treatment of wounds. Close attention to details is important in the technique of a first dressing after an operation. All instruments, irrigating fluids, bowls, basins, etc., are to be sterilized. When the dressing is removed the skin surrounding the wound should be cleansed by washing with salt solution or peroxide of hydrogen. The sutures or drainage should be removed with sterile forceps, and fresh sterile dressings applied. If a wound is found infected, all accumulations of blood-clot, pus, etc., should be gently and carefully washed out, and free drainage provided. In the care of infected wounds careful attention must be paid to maintaining mechanical cleanliness and avoiding infection with some organism which may not be already present.

It should be borne in mind that anything that tends to depress a patient's resisting powers encourages infection; such as prolonged exposure to cold during an operation, loss of blood and infliction of a great degree of surgical shock.

PART III.

NON-PATHOGENIC BACTERIA.

THE number of varieties of non-pathogenic bacteria is very large. Eisenberg[1] describes 376 species of bacteria, mostly non-pathogenic. Sternberg[2] enumerates 489 species, including the pathogenic varieties, but the majority, of course, are non-pathogenic. Flügge[3] considers about 500 species of bacteria. Migula[4] recognizes nearly 1,300, and Chester[5] about 800 species. Probably some of the bacteria which have been described as distinct species are in reality not different; but, on the other hand, it is also probable that a still larger number of species have not been described at all; how many it is impossible to say. In a work of this character it is feasible to mention only a few of the commonest and best-known species of non-pathogenic bacteria.

Micrococcus agilis.—Found in water; coccus about 1 μ in diameter, usually appearing as diplococci, sometimes as streptococci and tetrads; liquefies gelatin slowly; grows at room temperature, on ordinary culture-media, forming a rose-red pigment on agar and potato. This micrococcus is remarkable in being actively motile; it possesses a flagellum. It is stained by Gram's method.

Micrococcus ureæ.—Found in decomposed, ammoniacal urine and in the air; coccus .8 to 1 μ in diameter, occurring

[1] *"Bakteriologische Diagnostik,"* 1891.
[2] "Manual of Bacteriology," 1893.
[3] *"Die Mikroörganismen,"* 1896.
[4] *"System der Bakterien,"* 1900.
[5] "Manual of Determinative Bacteriology," 1901.

singly or in various combinations; does not liquefy gelatin; facultative anaërobic; grows rapidly, best at 30° to 33° C.; grows on ordinary gelatin, but best on special media; it decomposes urea, producing ammonia and carbon dioxide, which form ammonium carbonate.

Sarcinæ.—There is a large number of species of sarcinæ. They are common organisms in the air. They frequently contaminate plate-cultures. Many of the sarcinæ of the air present, in cultures, growths having brilliant colors, from which some of them are named; thus there are orange, yellow, rose-colored and white sarcinæ, and others.

Sarcina pulmonum.—Found in the air passages of man; 1 to 1.5 μ in diameter, occurring in tetrads or cubes of eight cells; aërobic; does not liquefy gelatin; grows slowly, best at ordinary temperatures, preferably upon gelatin. It decomposes urine with the formation of ammonia. It is said to form endogenous spores which are extremely resistant to heat.

Sarcina ventriculi.—Found in the stomachs of man and of animals; 2.5 μ in diameter, occurring in cubes of eight cells or more; it does not liquefy gelatin; aërobic; grows on ordinary culture-media; the growths tend to become yellow. Small numbers of sarcinæ may occur in the normal human stomach; the presence of large numbers indicates the existence of abnormal fermentative processes.

Bacillus fluorescens liquefaciens.—Found in water and putrid fluids; very common; appears as a small rod, actively motile; aërobic, but somewhat variably; liquefies gelatin; grows rapidly at ordinary temperatures upon the usual culture-media. It forms a pigment having a beautiful greenish-yellow fluorescence, best seen in transparent media; the growth on potato has a brown color. Does not stain by Gram's method and does not form spores.

Bacillus fluorescens putidus.—Found in water; a short rod with rounded ends; actively motile; does not liquefy gelatin; aërobic; does not form spores; grows rapidly at the ordinary temperatures upon the common media. Gelatin cultures give off a powerful, foul odor of trimethylamin. It produces a greenish, fluorescent pigment, best seen in transparent media; on potato the growths form a thin, gray to brown, slimy layer.

There are several other fluorescing bacilli, mostly found in water.

Bacillus Indicus.—Found by Koch in the stomach-contents of an ape in India; a fine short bacillus with rounded ends; motile; does not form spores; facultative anaërobic; liquefies gelatin; grows rapidly, best at 35° C. upon the ordinary media; produces a brick-red pigment. Very large doses injected into rabbits caused death in three to twenty-four hours.

Bacillus prodigiosus.—Widely disseminated in the atmosphere of certain places; a short bacillus with rounded ends, in form often nearly like the micrococci; facultative anaërobic; not motile, as a rule; does not form spores; liquefies gelatin rapidly; grows rapidly, best at 25° C. on the ordinary culture-media; milk is coagulated; gas forms in sugar-media; cultures on potatoes give off a foul odor of trimethylamin. A brilliant red color, which only develops in the presence of oxygen, appears in cultures. The pigment appears as granules outside of the bacteria.

Bacillus violaceus (of Berlin).—Found in water; a slim rod with rounded ends which may form threads; actively motile; facultative anaërobic; liquefies gelatin rapidly; forms endogenous spores placed near the centers of the bacilli; grows rapidly, and not at high temperatures, upon ordinary media, forming a deep, violet-colored pigment. There are several bacilli related to this one.

Bacillus amylobacter (Clostridium butyricum, Bacillus butyricus, Prazmowski).—Found widely distributed in nature in decomposing vegetable material and in the stomachs of ruminant animals; a large, thick rod with round ends, often arranged in chains; actively motile; anaërobic; forms spores, which are located in the center of the bacillus and give it a spindle-shaped form, or at one end when it has the outline of a tadpole; has not been cultivated satisfactorily on ordinary media; grows best at 35° to 40° C.; decomposes carbohydrates with the formation of butyric acid; decomposes cellulose. Organisms of similar form have been found as fossils belonging to the carboniferous period.

Bacillus butyricus (Hueppe).—Found in milk; appears as a small, irregular rod, also forming threads; very actively motile; aërobic; rapidly liquefies gelatin; forms centrally located spores; grows best at 35° to 40° C.; grows rapidly on ordinary media; coagulates milk, redissolving the coagulum, producing also butyric acid. A large number of bacteria, both aërobic and anaërobic, produce butyric acid fermentation.

Bacillus megatherium.—Obtained by DeBary from cooked cabbage-leaves; common on plants and earth; a large bacillus with rounded ends, often forming chains; motile; slowly liquefies gelatin; aërobic; forms spores, especially in potato cultures; grows rapidly at room temperature on the ordinary media.

Bacillus mesentericus vulgatus (Potato bacillus).—Found on potatoes; common in earth; a large, long rod with rounded ends, often forming long chains; motile; it is stained by Gram's method; liquefies gelatin; aërobic; forms spores; grows rapidly, best at about 20° C.; grows on ordinary media, forming on potato a thin, wrinkled membrane which spreads rapidly over the surface. It coagulates milk, redissolving the coagulum. It possesses numerous flagella. The spores are extremely resistant to heat.

Bacillus phosphorescens Indicus.—Obtained from sea-water; a small, thick, rod-shaped bacillus with rounded ends, also forming threads; actively motile; not stained by Gram's method; liquefies gelatin; aërobic. It grows slowly, best between 20° and 30° C., upon the usual media; except milk and potato. Its cultures, when old, especially when on animal nutrient-media and in the presence of certain sodium salts, are phosphorescent in the dark.

There are various other bacilli which produce phosphorescence, some of which do not liquefy gelatin.

Bacillus mycoides (Bacillus ramosus, *Wurzelbacillus*). Found in the earth and in water, very common; a large, short bacillus with rounded ends, often forming chains and threads; slightly motile; liquefies gelatin; aërobic; forms centrally located, oval spores; grows rapidly at room and incubator temperatures upon the usual media. It is said to rapidly decompose albumin with the formation of ammonia.

Bacillus subtilis (Hay bacillus).—Found on hay, in the air, water, ground and decomposing fluids; very common; a large bacillus somewhat resembling the anthrax bacillus in form, with rounded ends, often forming chains or long filaments; motile; possessing flagella; liquefies gelatin; aërobic; it is stained by Gram's method. It may have large, centrally located spores, which form best on potato at about 30° C. The spores are extremely resistant to heat and to chemical germicides. It grows best at about 30° C. upon the ordinary culture-media; milk is peptonized. Bacillus subtilis may easily be isolated in pure culture by adding finely-cut hay to bouillon; place in the steam sterilizer for five or ten minutes; then let the tubes develop in the incubator. Plates made from the bouillon will probably show colonies of the bacillus subtilis only, as the steam may be expected to have destroyed all organisms except its very resistant spores. The hay bacillus is entirely without pathogenic properties.

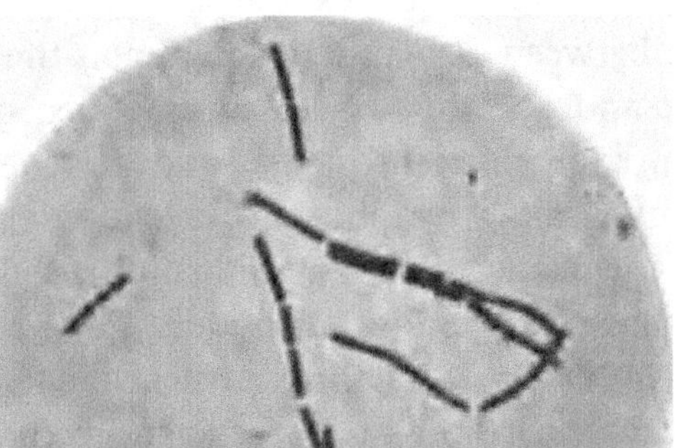

Bacillus subtilis.   ($\times$ 1000.)

Bacillus erythrosporus.—Found in decomposing fluids and water; a slim bacillus with rounded ends; motile; does not liquefy gelatin; facultative anaërobic; forms oval, red-colored spores, two to eight in each filament; grows rapidly, only at ordinary temperatures; produces a greenish-yellow fluorescent pigment. On potato it forms a limited, reddish growth, becoming nut-brown.

Bacillus cyanogenus (Bacterium syncyanum, Bacillus lactis cyanogenus, Bacillus of blue milk).—A bacillus of variable size, with rounded ends; motile; spore formation doubtful; is aërobic; not stained by Gram's method; grows rapidly at ordinary but not so well as incubator temperatures on the usual culture-media; does not liquefy gelatin; produces a grayish-blue pigment, brighter in acid media, at ordinary temperatures; milk is not coagulated, or rendered acid.

Bacillus acidi lactici (Hueppe).—Found in sour milk; a short, plump rod; not motile; does not liquefy gelatin; facultative anaërobic; grows on the ordinary media; in milk causes development of lactic acid with precipitation of casein and production of gas and alcohol. It belongs in the same group as B. coli communis and B. lactis aërogenes (see Part IV.).

There are numerous other bacteria, such as the bacterium acidi lactici, which cause the formation of lactic acid in milk.

Bacterium ureæ.—A short, thick bacillus with rounded ends; not motile; aërobic; found in ammoniacal urine; grows slowly at room temperature upon gelatin, which is not liquefied; decomposes urea, forms ammonium carbonate.

Bacterium Zopfii.—Found in the intestines of hens, in water and in fecal matter; a bacillus .75 to 1 μ broad and 2 to 5 μ long; may form threads. Actively motile; does not liquefy gelatin; aërobic; involution forms are often seen and they have been described as spores; grows rapidly, best at 20° C. upon gelatin; forms branching zoöglœæ. It is a member of the same group as B. proteus (see Part IV.).

Spirillum rubrum.—Found by Esmarch in the putrefying cadaver of a mouse; short spirals twice the breadth of the cholera spirillum, usually with one to three turns; in bouillon growing into long spirals; motile with flagella; spore formation doubtful; facultative anaërobic; does not liquefy gelatin; grows slowly, best at about 37° C. on the ordinary media; produces a wine-red pigment only when the air is excluded.

Spirillum or Spirochæta dentium.—Found in the mouths of healthy persons, on the margins of the gums when they are covered with a dirty deposit; long spirals with several windings uneven in thickness; has not been cultivated.

Spirillum sputigenum.—Found in the human mouth in healthy persons at the margin of the gums; curved rods or short spirals which resemble the spirillum of cholera in form; has not been cultivated.

Spirillum rugula (Vibrio rugula).—Found in swamp water, in fecal matter, and in the tartar of the teeth; a curved rod .5 to 2.5 μ broad and 6 to 8 μ long, having one flat spiral winding; motile, with flagella at the ends; probably anaërobic; forms spores located at the ends.

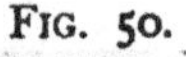

FIG. 50.

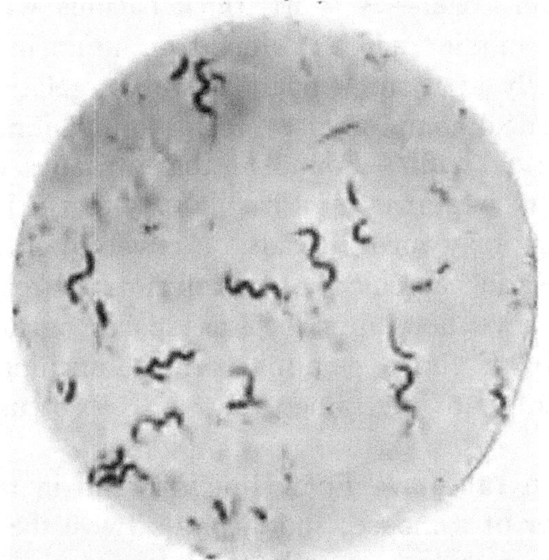

Spirilla from Swamp Water.   ($\times$ about 500.)

Spirillum volutans.—Found in swamp water; very long spirals with several turns; 1.5 to 2 μ broad and 25 to 30 μ long; motile, with a flagellum at each extremity.   The protoplasm is granular.

Spirillum undula.—Found in putrefying infusions containing organic matter; a rather short spiral form with three turns or less, about 1 μ thick and 8 to 12 μ long; actively motile, with a tuft of flagella at each extremity; has been cultivated on agar.

Spirillum or Spirochæta plicatile.—Found in swamp water; spiral forms of various lengths; sometimes 100 to 200 μ long; actively motile.

The spirilla (vibrios or comma-shaped forms), closely resembling the spirillum of cholera, will be considered in connection with that organism. A form of chronic pseudo-

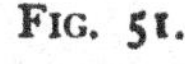

FIG. 51.

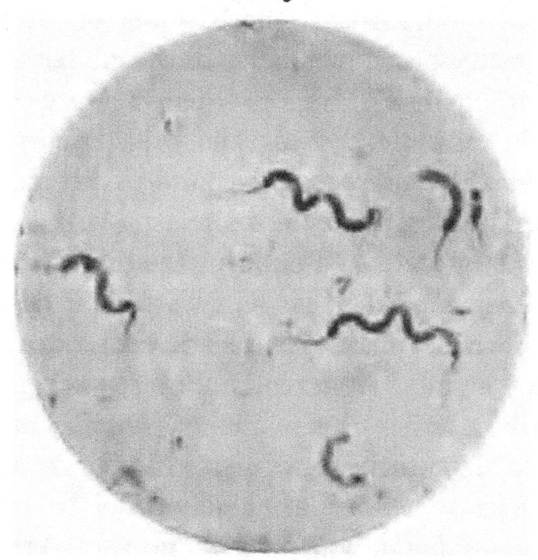

Spirilla from Swamp Water, showing Flagella, Löffler Stain. (× 1000.)

membranous inflammation of the pharynx has been attributed to an organism called the fusiform bacillus or spirillum of Vincent.[1]

Higher Bacteria.—Certain organisms (beggiatoa, thiothrix, leptothrix, cladothrix, actinomyces or streptothrix) of more complicated structure than most bacteria, but resembling them in many respects, are called " higher bacteria." They consist of definite filaments which are usually made up of rod-shaped elements, but the relation between these elements is very intimate. Some of them (beggiatoa,

[1] Mayer, *American Journal Medical Sciences,* Vol. 123, 1902, p. 187.

thiothrix) contain sulphur granules. Many of them occur in water. There are forms among them which are found attached to some object by one end of the filament (thiothrix). Some of them (actinomyces or streptothrix) have branching filaments, which are rarely seen among the lower bacteria (see page 119). Often one end of the filament becomes specialized for the purposes of reproduction. The fungus of actinomycosis is the best known of this group. There are many other members, however, both pathogenic and non-pathogenic. Most of them require still further study. The tubercle bacillus and other acid-proof bacilli which resemble it, have some points of resemblance with actinomyces (see B. tuberculosis, Part IV.).

Leptothrix buccalis.—Found in the mouth cavity. This name has been applied to large, twisted, thread-like organisms, in which segments can be demonstrated with difficulty or not at all. Apparently, different organisms have been described under this name. Vignal claims to have cultivated a leptothrix buccalis. Miller recognizes two principal species, neither of which could be cultivated,— *leptothrix innominata,* which shows no transverse divisions, and which is stained faintly yellow by iodine; and *bacillus buccalis maximus,* in which the transverse divisions are distinct, and which is stained brownish-violet by iodine. Miller's *leptothrix maxima buccalis* is similar to the last except in lacking the iodine reaction.

A variety of leptothrix, or a nearly related organism, appears to be the most frequent cause of the form of gangrenous inflammation of the mouth and genitals called noma. It stains faintly by Gram's method. It does not grow on ordinary media.[1] Another organism of this group has been

---

[1] Blumer and MacFarlane, *American Journal Medical Sciences,* November, 1901.

described which is pathogenic to a number of domestic animals.[1]

Yeasts and Moulds.—In the course of bacteriological work one constantly encounters yeasts and moulds, which, although not bacteria, must nevertheless be understood and recognized to avoid error. Accidental contamination of tubes or plates is likely to be the result of the growth of some

FIG. 52.

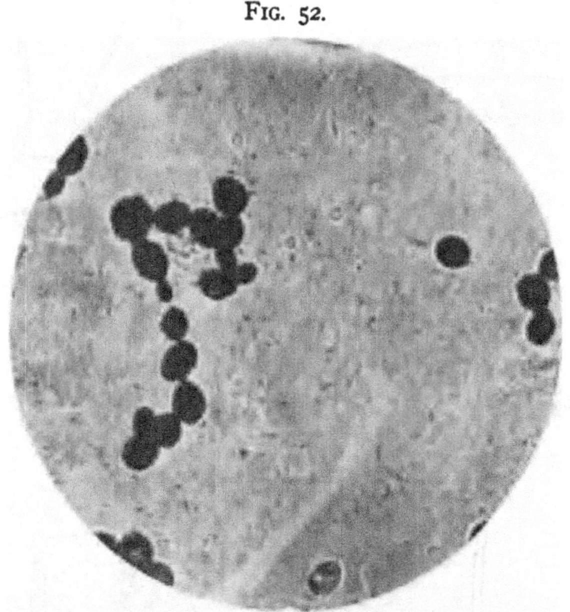

Yeast Cells, stained with Fuchsin. (× 1000.)

of these forms. The yeasts generally go by the name of *saccharomyces*, of which there are several species. The *saccharomyces cerevisiæ* is the ordinary yeast of alcoholic fermentation. Some of the yeasts present colored growths —red, white and black. They consist of large, oval cells, which readily stain with the aniline dyes. They multiply by

[1] It has also been called "Necrosis bacillus," and "Streptothrix cuniculi." Pearce, *University of Pennsylvania Medical Bulletin*, November, 1902.

FIG. 53.

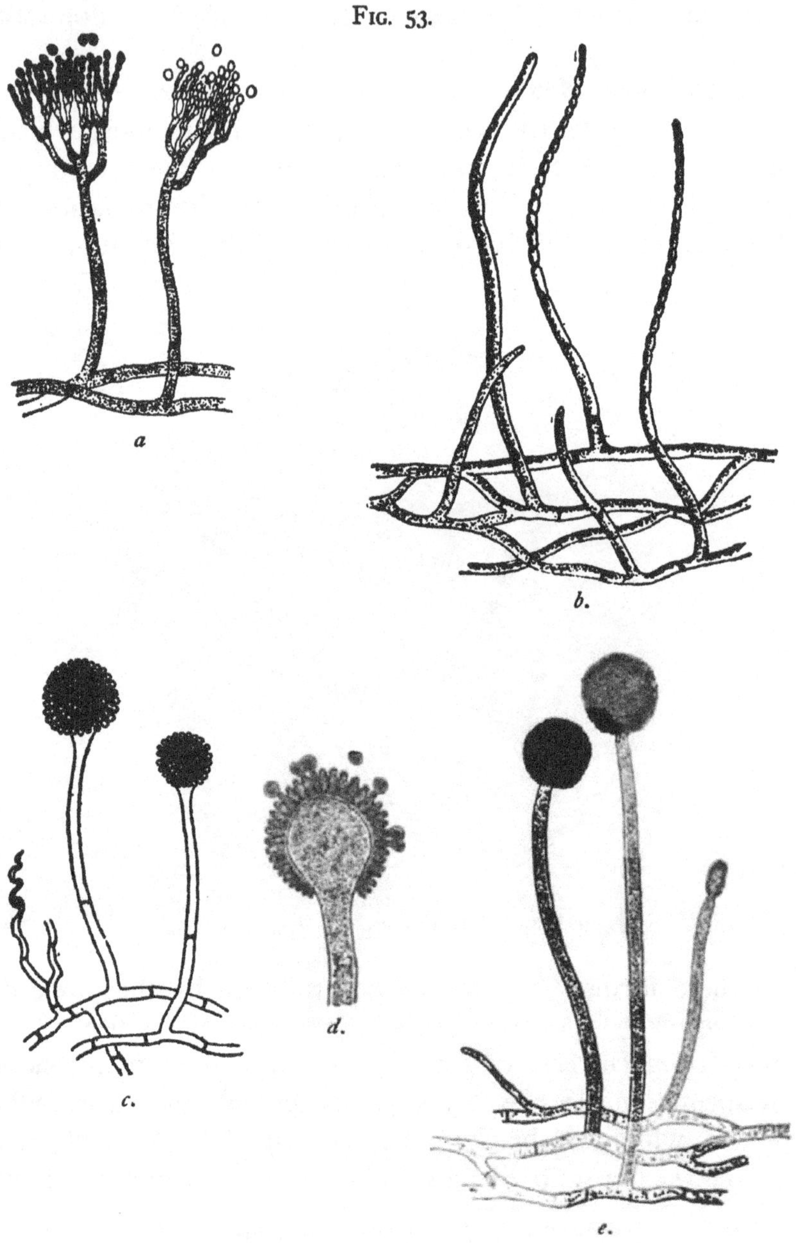

a. Penicillium glaucum. b. Oidium lactis. c. Aspergillus glaucus. d. The
same more highly magnified. e. Mucor mucedo (Baumgarten).

the protrusion of a little bud from the cell, which develops into a new cell. In an actively germinating growth of yeast these budding cells are readily distinguished (Fig. 52).

Yeasts have been found that were pathogenic to animals. They have also been supposed to be the cause of some malignant tumors, but this view has been, for the most part, abandoned.

Among the moulds the varieties most commonly encountered are the *mucor,* the *penicillium,* the *aspergillus* and the *oidium.* There are various species of each of them. They consist of cells arranged end to end, making a thread-like body called a *hypha.* The threads are matted together and form a *mycelium.* Certain threads project upward from the mycelium, and on them are borne *spores,* or conidia. The arrangement of the spores is characteristic in each variety of mould (Fig. 53). A group of organisms exist which have affinities both with yeasts and mould-fungi. Some of them are pathogenic. The form of infection of the mouth called thrush, is due to a fungus of this class, which is generally considered an oidium. A chronic inflammatory affection of the skin (blastomycetic dermatitis) is due to related organisms.[1] The Sporotricha of Schenck[2] which produces chronic subcutaneous abscesses, may be mentioned here, provisionally. A number of skin affections, such as Tinea favosa and Tinea trichophytina, are due to fungi, which have some similarity with those above mentioned.

Among the mould fungi, several species of aspergillus and of mucor are pathogenic. Man, as well as the lower animals, may be affected. In man the lungs may be involved in a broncho-pneumonia (pneumonomycosis), usually due to

[1] Ricketts, *Journal of Medical Research,* Vol. VI., 1901; Hyde and Montgomery, *Journal American Medical Association,* June 7, 1902.

[2] Hektoen, *Journal Experimental Medicine,* Vol. V.

aspergillus, and often secondary to some preëxisting disease of the lung. Mould fungi, especially aspergillus, may grow in the external ear (otomycosis). The growth is usually superficial. These fungi rarely produce lesions in other organs.

PART IV.

Suppuration and Allied Conditions.—The occurrence of suppuration is characterized by certain appearances which we are accustomed to describe under the name of inflammation. The study of inflammation belongs to pathology, and cannot be considered here. However, certain evidences which are characteristic of the suppurative variety of inflammation need to be outlined on account of their relation to the action of the pyogenic bacteria.

In a suppurating area, as is well known, the blood-vessels are dilated, and the lymph-spaces become filled with serum. Leucocytes are attracted to the neighborhood in large numbers, we may suppose by a positive chemotaxis, and crowd the small veins and capillaries. The leucocytes, by reason of their amœboid movement, pass through the walls of the vessels at little openings filled with cement-substance, situated between the lining endothelial cells. According to the theory of phagocytosis, they are bent on finding the irritant which has led to the inflammation, and upon isolating it and rendering it harmless. At the point which appears to be the center of the inflammatory area there is usually, but not always, a necrosis of the cells of the tissue; this constitutes the central slough or the familiar core of some boils. The necrosis is to be attributed to poisons formed by the micrococci. In sections cut through such an abscess the nuclei of the central necrotic cells fail to take the nuclear stain; the necrotic mass does not stain, or takes the dye diffusely and irregularly, and it exhibits many fine granules.

We find the cells of the tissues surrounding the necrotic area mingled with large numbers of polynuclear leucocytes, which enclose the area of irritation.

The nuclei of the cells near the center of the abscess are frequently broken up into a number of small parts (fragmentation), which indicates the commencement of their destruction. In sections through small abscesses it is possible, by means of a double stain of carmine followed with gentian-violet, according to Gram's method, to bring out the histological character of the tissue, and at the same time to stain the common pyogenic bacteria, which are usually found near the center of the abscess in large numbers, even making masses visible with a low power of the microscope. Preparations, most convincing and of great beauty, may be secured in this manner. It is often possible to demonstrate masses of micrococci filling up the lumina of capillaries in which they are lodged as emboli.

The production of pus in the center of the abscess is due to the liquefaction of the necrotic tissue, which apparently results from the action of some peptonizing ferment. In the liquid thus formed, immense numbers of the polynuclear leucocytes are found floating, and they constitute the greater part of the so-called *pus-cells*. The nuclei of these cells are obscured by clouds of extremely fine granules. The granules are of an albuminoid nature, and are dissolved by acetic acid, when the nuclei become visible. The nuclei generally consist of three, four, five or more portions. The presence of the fine albuminoid granules in the pus-cells is to be counted as a degenerative change. Although it is possible to produce suppuration in laboratory experiments by the introduction of sterilized irritants, such as croton oil, in the vast majority of cases suppuration is due to the action of pyogenic bacteria.

Specimens of pus will nearly always be found to contain bacteria, which can be demonstrated by cultivation, and, as a rule, also in smears made and stained upon cover-glasses. The bacteria are generally found outside the pus-cells. In the case of the gonococcus and the diplococcus intracellularis meningitidis they are characteristically found in pairs, inside of, or at least attached to the pus-cells. The character of the suppuration differs somewhat with the different species of pyogenic bacteria. The kind of abscess above described—localized and having a central slough, usually rather slow in progress—is typical for the staphylococcus pyogenes aureus, which is prone to produce circumscribed areas of suppuration. The streptococcus pyogenes, on the other hand, oftener leads to suppuration of a more diffused character, such as we see in cellulitis and erysipelas; but either organism may, at times, produce the effects usually characteristic of the other. Pus having a blue or green tinge generally owes the color to the presence of the bacillus pyocyaneus. The commonest pus-producing organism is then the *staphylococcus pyogenes aureus,* and next to that the *streptococcus pyogenes.* Among the other pyogenic bacteria the following may be named:

Staphylococcus pyogenes albus, including staphylococcus epidermidis albus; streptococcus of erysipelas (probably identical with streptococcus pyogenes); gonococcus; diplococcus intracellularis meningitidis; staphylococcus pyogenes citreus; micrococcus tetragenus; micrococcus pyogenes tenuis, which may be the same as the micrococcus lanceolatus; staphylococcus cereus albus and flavus.

Pus-formation may also be due to micrococcus lanceolatus, bacillus pyocyaneus, bacillus proteus, bacillus coli communis, bacillus pyogenes fetidus, bacillus pneumoniæ (of Friedländer), bacillus aërogenes capsulatus, the ray fungus of actinomycosis, and possibly the bacillus of bubonic plague. Besides these organisms, there are others whose effects are usually more marked in a specific way which sometimes form pus, as the bacilli of diphtheria, tuberculosis, glanders and typhoid fever.

Frequently two or more species of pyogenic bacteria will be found associated.

The table on page 239, quoted from Dowd, shows the frequency of the occurrence of various pyogenic bacteria in 135 cases of different types of suppuration.

The condition of the animal's tissues is of great importance in determining whether or not suppuration is to occur. It will be seen that we are repeatedly subjected to infection with pyogenic bacteria, but that in most cases suppuration nevertheless does not occur. The local conditions have an important influence in determining infection. Regions of hyperemia, edema, anemia or necrosis are especially liable to suppuration, as are tissues which have been bruised, lacerated, strangulated or otherwise damaged. Furthermore, the general condition of the patient is of great importance. Chronic diseases and conditions of exhaustion or depression dispose to suppuration, and the depraved condition of the tissues in diabetes renders the sufferer from this disease especially liable to it. These facts have already been enumerated in a previous chapter (page 165). In the lower animals we find that it is often very difficult to produce suppuration artificially with the ordinary pyogenic bacteria. In rabbits the subcutaneous introduction of staphylococcus pyogenes aureus frequently fails to produce an abscess. Suppuration is likely to result, however, if an irritant body like a piece of sterilized potato or sterilized glass be introduced along with the bacteria.

Pyogenic bacteria are most frequently introduced into the body through the agency of injuries and wounds of various sorts. They are very widely disseminated in nature, and are always liable to be clinging to external objects, especially in cities and around dwellings. The infection of a wound in this manner, when the suppuration is of a spreading character, such as is most characteristic of streptococcus infection, is known in every-day lan-

	Cellulitis, 51 Cases.	Infected Fresh Wounds, 17 Cases.	Old Granulating Wounds, 18 Cases.	Healing Wounds: Stitches, 5 Cases.	Furuncles, 7 Cases	Abscesses, 37 Cases.
Streptococcus pyogenes alone	9	3				8
Streptococcus pyogenes predominant	23	3				8
Streptococcus pyogenes relatively few	3	1	6			1
Staphylococcus pyogenes aureus alone	11	1	1	1	7	6
Staphylococcus pyogenes aureus predominant	8	2				1
Staphylococcus pyogenes aureus relatively few	13	3				2
Staphylococcus pyogenes or epidermidis albus alone	1	4	2	4		2
Staphylococcus pyogenes or epidermidis albus predominant		1				
Staphylococcus pyogenes or epidermidis albus relatively few	10	5	3			6
Staphylococcus cereus albus	3	1	2			1
Staphylococcus citreus	1		2			1
No growths on agar						11
Very few growths on agar			3			3
Bacillus pyocyaneus			1			3
Bacillus coli communis						3
Overgrown	4		2			1
Few undetermined colonies	12	2	5			5

guage as " blood-poisoning." It is possible for infection to take place around hair-follicles through the unbroken skin. In such instances the suppurative inflammation first shows itself in a minute red pimple with a hair in the center. The pimple presently becomes a pustule. The process may cease at this point, or it may be only the commencement of a large carbuncle with a central slough. Such infection has been produced experimentally on the human skin by rubbing in cultures of staphylococcus pyogenes aureus. It is, furthermore, the constant experience of post-mortem examiners that infection may occur around the hair-follicles when no wound of the skin has been inflicted.

In many instances, infection with the pyogenic bacteria follows upon some preëxisting infection; this happens, for instance, in tuberculosis, when tuberculous lungs become infected with streptococcus pyogenes, leading to the formation of a cavity. It is a common occurrence in gonorrhea, after the acute stage of the disease has passed, when we find the gonococcus in the pus, mingled with other pyogenic micrococci. Secondary infection with pyogenic bacteria is frequently due to the streptococcus pyogenes, often also to the micrococcus lanceolatus.

Sometimes we are obliged to admit that the manner in which the pyogenic bacteria enter the body is unknown.

The severe general symptoms, familiar to every physician, often accompanying acute suppuration, indicate the formation of toxic bacterial products and their absorption. Experimental evidence of the formation of such toxic products is not so clear, however, for the pyogenic organisms as for some of the other bacteria. It has been shown that cultures of staphylococcus pyogenes aureus, in which the bacteria have been killed, are capable of producing suppuration in the lower animals.

The pyogenic bacteria play a somewhat different part in producing disease, which is fully as important as the typical suppuration seen in an abscess. This happens when the suppurative condition is mixed with other phenomena, or when there is inflammation of another variety without suppuration at all; or there may be lesions not inflammatory in a strict sense. These differences in their action depend largely upon the organ affected. One such condition is osteomyelitis, which is, usually, suppuration occurring in bone, but which does not present the ordinary picture of pus-formation owing to the hard and unyielding character of the tissue. Other conditions of very great importance are meningitis, pericarditis, pleuritis, pneumonia (croupous and broncho-), peritonitis and endocarditis. It will be observed that these affections are, for the most part, inflammations of the serous membranes. Such inflammations, when they are produced by pyogenic bacteria, are likely to be of great severity, accompanied by the formation of fibrinous exudates; pus-formation may or may not be present. We find that the cause at times is the staphylococcus pyogenes aureus; this is often the case in malignant endocarditis. Generally speaking, in such inflammations the streptococcus pyogenes, the staphylococcus pyogenes aureus, and the pneumococcus occur most commonly, although they are by no means the only organisms found. Many cases of peritonitis show the presence of B. coli communis, either in combination with other bacteria, or alone.[1] This is explained by the proximity of the intestine, and especially by the frequent occurrence of peritonitis after perforation of the intestine.

In inflammations of mucous membranes the common pyogenic organisms play the most important though not an exclusive part. In acute bronchitis, pneumococci and streptococci were found by Ritchie to be the commonest causes.

In inflammations of the middle ear the principal causes are the pneumococcus, the streptococcus, and the staphylococcus aureus and albus.[3]

In 25 cases of acute cystitis in women Brown[4] found B. coli communis, 15 times; S. pyogenes albus, 5 times; S.

[1] Flexner, "Etiology, etc., of Peritonitis," *Philadelphia Medical Journal*, November 12, 1898.

[2] Ritchie, *Journal Pathology and Bacteriology*, Vol. VII., December, 1900.

[3] Hasslauer, *Centralblatt f. Bakteriologie*, XXXII., Ref., 1902, p. 174. Compare *Ibidem*, pp. 240 and 246.

[4] *Johns Hopkins Hospital Reports*, Vol. X., 1902.

pyogenes aureus, 2 times; B. typhosus, 1 time; B. pyocyaneus, 1 time; B. proteus vulgaris, 1 time.

A number of investigators have recovered from cases of acute articular rheumatism organisms resembling the pyogenic cocci. Most frequently a diplococcus or short streptococcus has been found, which has sometimes produced arthritis and endocarditis when inoculated into rabbits.

FIG. 54.

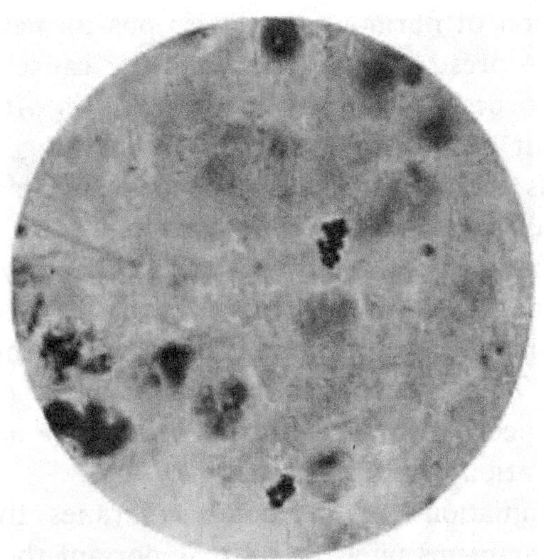

Staphylococcus pyogenes aureus in Pus, stained by Gram's Method.
(× 1000.)

From a point where there is suppuration or other localized infection, pyogenic bacteria may enter the circulation and become widely disseminated throughout the body. That happens very commonly in malignant endocarditis. In this manner secondary or metastatic abscesses may be produced in the most diverse organs.

The term *pyemia* is used to describe the dissemination of pyogenic bacteria in the circulating blood, with the formation of metastatic abscesses.

Staphylococcus pyogenes aureus.—A micrococcus of variable size, arranged in irregular clumps, sometimes in pairs; about .8 to .9 μ in diameter; not motile (Fig. 55). It stains by Gram's method; it is facultative anaërobic; grows rapidly, best at 30° to 37° C. It liquefies gelatin. Upon gelatin plates small colonies appear at the end of about two days. It grows well upon all the culture-media. Milk is coagulated. It does not lead to fermentation with the production of gas but produces various acids.

FIG. 55.

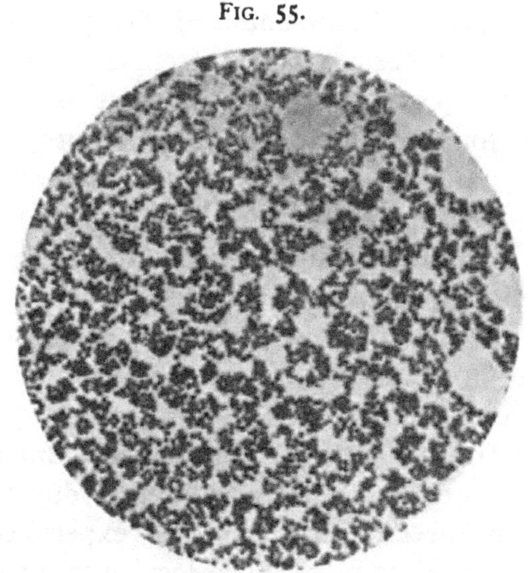

Staphylococcus pyogenes aureus, Pure Culture. (× 1000.)

The growths in the first place are pale, subsequently becoming golden-yellow in color, but only in the presence of oxygen. This color appears well on all media, and is especially distinct on potato. Sometimes the color is slow in developing.

In a fresh, moist condition the organism is killed by ten minutes' exposure to 58° C.; in a desiccated condition it requires a temperature of 90° to 100° C. to destroy it. It resists drying in a considerable degree. In the same specimen the micrococci may have quite different resisting powers to chemical germicides. Some of them are destroyed by 1–1000 solution of bichloride of mercury in five minutes; others survive exposure to the same for from ten to thirty minutes. (Abbott.)

Sterilized cultures introduced into animals may produce local suppuration. The toxic substances occur in the bacterial cells.[1]

As has already been mentioned, the staphylococcus pyogenes aureus is the commonest of the pyogenic bacteria in man. It has been obtained from a great variety of sources, and appears to be able to exist as a saprophyte. It has been found on the skin, in the mouth, in the nasal and pharyngeal mucus, and also in the alimentary canal. It has furthermore been detected in the air and in dust. It appears to find the conditions necessary for its existence in the vicinity of human habitations.

Cultures of the staphylococcus pyogenes aureus vary considerably in virulence. These variations are sometimes to be explained through cultivation on unfavorable media or repeated transplantation from one medium to another; but at times the diminished virulence is due to unknown causes. The lower animals used for experiments are not as readily infected as man. The local introduction in rabbits or guinea-pigs of a part of a culture of staphylococcus pyogenes aureus may be entirely without effect. The use of a very large dose, or the addition at the same time of some kind of irritant, may produce an abscess. Large amounts of cultures in bouillon may often be injected into the peritoneal cavity of the dog without effect, when the simultaneous addition of a piece of sterile potato or an injury to the gut may lead to fatal peritonitis. Introduc-

---

[1] See also Morse, *Journal Experimental Medicine*, Vol. I., p. 613.

tion of fluid cultures into the venous circulation of the rabbit generally produces metastatic abscesses in the kidneys, the heart-muscle and the voluntary muscles, and causes death.

In man this organism produces suppuration of a localized character, such as we are familiar with in boils and carbuncles. It has been shown to be the usual cause of infectious osteomyelitis. Osteomyelitis has been produced experimentally in rabbits by the injection of the staphylococcus pyogenes aureus, both with and without previous injury to the bone of the animal. Ulcerative endocarditis has on numerous occasions been shown to be due to this organism. It has been found possible to produce ulcerative endocarditis experimentally in animals by the injection of the staphylococcus pyogenes aureus when the valves of the heart have first been mechanically injured. The staphylococcus pyogenes aureus has also been found in acute abscesses of the lymph-nodes, tonsils, parotid gland, and mammary gland, in suppurating joint affections and empyema. It appears, furthermore, in acute inflammation of the serous membranes,—pleuritis, pericarditis, peritonitis,—although less frequently than the streptococcus pyogenes.

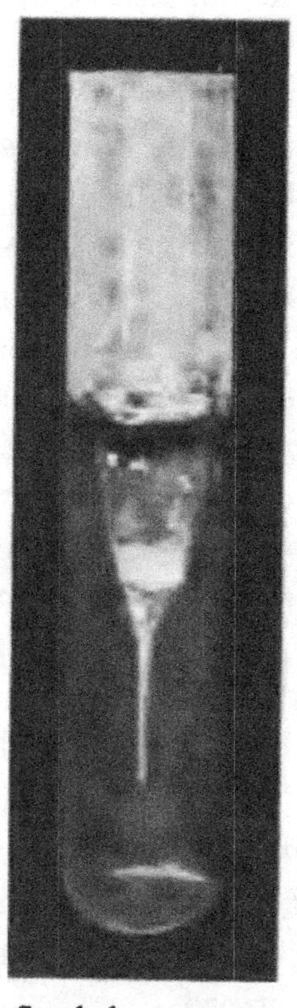

FIG. 56.

Staphylococcus pyogenes aureus, Gelatin Culture, 1 Week Old.

Staphylococcus pyogenes albus.—In form and manner of growth this organism behaves like the staphylococcus pyogenes aureus, with the exception that it produces no colored growths and its cultures appear white. Its pathogenic properties are less marked, and it is a less frequent cause of suppuration than the staphylococcus pyogenes aureus. It has, however, been found in acute abscesses on numerous occasions.

Staphylococcus epidermidis albus.—According to Welch, the epidermis of man contains with great regularity the organism to which he gave the above name, and which he considers to be a variety of staphylococcus pyogenes albus. It grows, liquefies gelatin, and coagulates milk more slowly than the ordinary staphylococcus pyogenes albus. It is, furthermore, possessed of less marked pus-producing tendencies. Welch found it impossible to sterilize the skin so as to remove this micrococcus from it. The organism is usually innocuous. It has been found in healthy wounds on numerous occasions. It is capable of causing trouble in wounds when necrotic or strangulated tissues are present, or where a foreign body like a drainage-tube has been left in the wound. It is a common cause of stitch abscesses.

Streptococcus pyogenes.—Appears as micrococci arranged in chains, often in pairs, when the adjacent cocci may be flattened. Sometimes the chains are very long. The diameters of the cocci vary from .4 to 1 μ. Attempts have been made to create varieties of streptococci according to the length of the chains. On that basis a streptococcus brevis and a streptococcus longus have been described.

The streptococcus pyogenes is not motile. It stains by Gram's method. By the method of Hiss (page 57) capsules may sometimes be demonstrated. It is facultative anaërobic; grows best in the incubator; more slowly at room temperature, and does not liquefy gelatin. In gelatin plates it produces small, round, white, punctiform colonies which are slow of development, and are only visible after about three days. It grows on the ordinary media; with the exception of potato, according to some authors. Milk may or may not be coagulated. The growths are never very luxuriant, and may die out entirely after a few transplantations.

It is killed by exposure to 52° to 54° C., in ten minutes. The streptococcus pyogenes occurs frequently on the mucous surfaces of the healthy body. It is often found in pus, especially pus of spreading inflammations of the kind

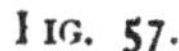

FIG. 57.

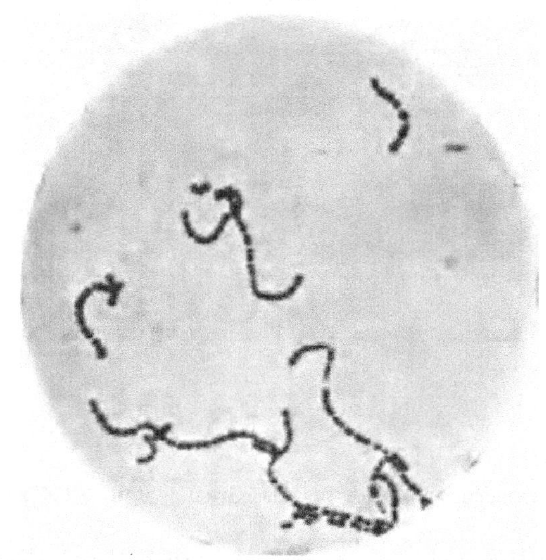

Streptococcus pyogenes, from a Pure Culture. (× 1000.)

known as cellulitis. This organism is the commonest infectious agent in puerperal fever, metritis and peritonitis. It occurs commonly in inflammations of the serous membranes —pleuritis, pericarditis and peritonitis. It has been discovered many times in ulcerative endocarditis, and in broncho-pneumonia. It is frequently present in the false membrane found in genuine diphtheria. It is also the cause of many of the pseudo-membranous or so-called " diphtheritic " affections of the throat where the Klebs-Löffler bacillus of diphtheria is wanting. These cases may be indistinguishable clinically from genuine diphtheria, and their nature will only be revealed on bacteriological examination. They are, however, as a rule, milder than genuine diphtheria. The pseudo-membranous affections of the throat which occur in scarlet fever and measles are generally caused

FIG. 58.

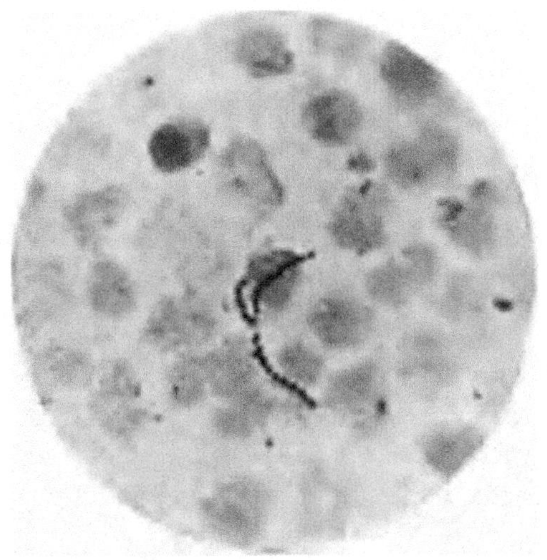

Streptococcus pyogenes in Pus, Gram's Stain. ($\times$ 1000.)

by the streptococcus pyogenes, although those diseases may be complicated by genuine diphtheria. Streptococci are very commonly present in the throat in scarlet fever,[1] and sometimes occur in the blood. Some observers believe that scarlet fever is caused by streptococci. Streptococci are very often found in the pustules of small-pox, and may also appear in the blood.

The streptococcus pyogenes is pathogenic for mice and rabbits, but the virulence is very variable. That may sometimes be increased by passing through a number of animals in succession, but is rapidly lost in artificial cultures. It is said that the virulence is best maintained when cultures on gelatin, after forty-eight hours' growth, are kept in a cool

[1] Weaver, *American Medicine,* April 18, 1903.

place, as in the ice-chest. Marmorek undertakes to. maintain or increase the virulence by growing it first in a mixture of human blood-serum (or that of the ass or the horse) with bouillon, and then inoculating it into the body of a rabbit, alternating these procedures, to obtain a culture of very high virulence. A serum of uncertain value derived from an immunized horse or ass and intended to cure streptococcus infection, has been prepared by Marmorek.

A number of other sera have been prepared to combat streptococcus infection. These have been used for human cases, including also scarlet fever. The results appear somewhat encouraging, although still uncertain.

It is said that streptococci may be agglutinated by serum from animals immunized to the streptococcus.

Coley has recommended a bouillon culture of streptococcus pyogenes (or of erysipelas), in which the bacillus prodigiosus was afterward grown, to be administered by injection, after sterilization of the cultures by heat, in cases of inoperable sarcomatous tumors. These injections appear in some cases to have accomplished remarkable and wholly unexplainable cures.

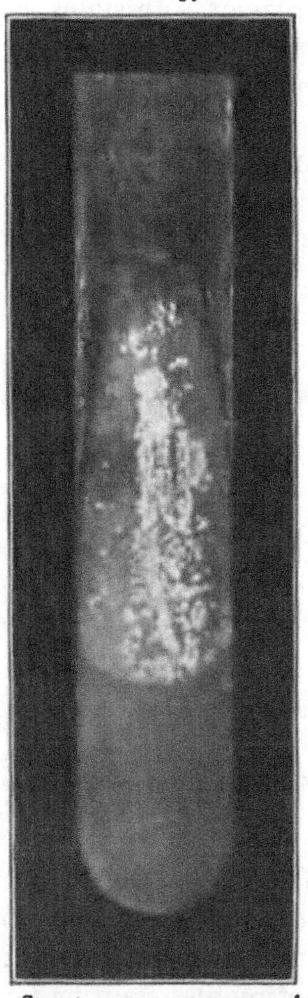

Fig. 59.

Streptococcus pyogenes, culture on agar (slightly enlarged).

Streptococcus of Erysipelas.—A streptococcus has been derived from cases of erysipelas which in all essential respects, in its morphology, its growth on culture-media, its behavior with stains, and its pathogenic properties, is similar to the streptococcus pyogenes. It is probable that these organisms are identical.

Micrococcus tetragenus.—Found in the cavities in the lungs of pulmonary tuberculosis, in sputum and in pus.

FIG. 60.

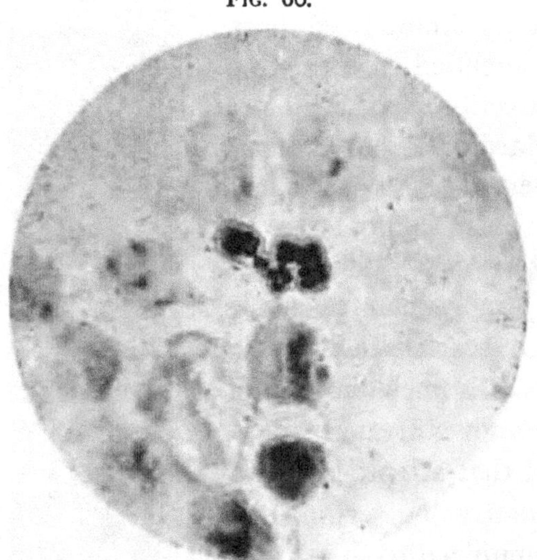

Micrococcus tetragenus in pus from a large abscess on the arm; showing capsule, Gram's stain and eosin.　(× 1000.)

The micrococci are enclosed in a transparent capsule, best seen in preparations from the tissues of inoculated animals, and are arranged in pairs or in fours; about 1 μ in diameter; not motile; stain by Gram's method. It grows well at the room temperature, but rather slowly; is facultative anaërobic; does not liquefy gelatin. Gelatin plates show little, white, punctiform colonies, which, with the low power, are finely granular, and have a peculiar glassy shimmer; and in stab-cultures the growths appear as little colonies along the line of puncture. On agar, round white colonies form, not spreading. It produces a thick, slimy film on potato and a broad, white, moist growth on blood-serum. This organism is only occasionally found in pus. It is pathogenic to white mice and guinea-pigs, not to gray mice and rabbits. It may produce a septicemia or only a localized suppuration in guinea-pigs. In white mice a general septicemia results, when the micrococcus tetragenus is found in the blood and in the great viscera. White mice usually die in from two to six days; guinea-pigs in from four to eight days.

Micrococcus lanceolatus (Micrococcus pneumoniæ crouposæ, Micrococcus Pasteuri, Diplococcus pneumoniæ, Micrococcus of sputum septicemia, Streptococcus lanceolatus Pasteuri, and Pneumococcus of Fränkel).—This organism was discovered by Sternberg in his saliva in 1880, and afterward demonstrated to be the cause of lobar pneumonia by Fränkel and Weichselbaum. The micrococci usually occur in pairs. The pair of micrococci, in its most typical form, appears like a couple of curved triangles with their bases close to each other. The outline is usually described as being lancet-shaped. The micrococci are frequently oval or round; they often form chains. When it is most characteristic, each pair of micrococci is surrounded with a capsule, which is best shown in preparations made from the blood of infected animals or from pneumonic sputum; the capsule is not usually seen in preparations made from cultures. For methods of demonstrating the capsule see page 57. The pneumococcus is not motile. It stains by Gram's method, which also is useful in demonstrating the capsule. It is facultative anaërobic. It grows only at elevated temperatures, preferably about 35° to 37° C. Gelatin is not liquefied. It grows well upon agar, upon blood-serum and upon Guarnieri's medium (p. 81). It does not grow upon potato. Milk usually becomes acid, and may or may not be coagu-

22

lated. The colonies are seen in their characteristic form upon agar, and are developed after about forty-eight hours, appearing as minute, whitish, translucent, circular growths.

It is killed by an exposure to 52° C. for ten minutes.

It is best cultivated from the blood of an animal which has been infected with the sputum of a case of lobar pneu-

FIG. 61.

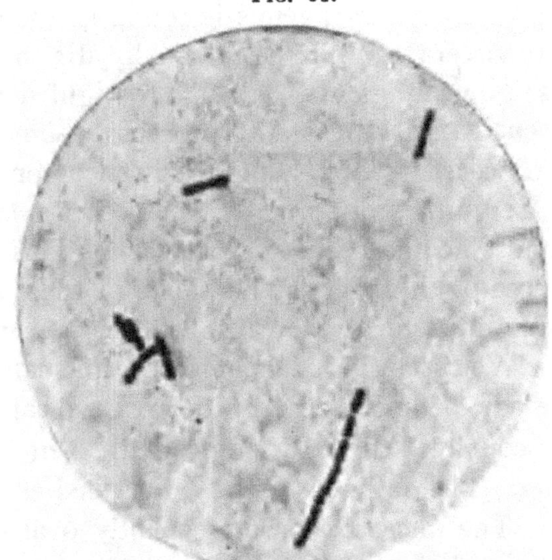

Pneumococcus of Fränkel in sputum of pneumonia, Gram's stain and eosin.   (× 1000.)

monia.   Cultures need to be transplanted every few days; they cannot usually be propagated more than a couple of months.

The virulence of the organism for animals diminishes rapidly in cultures.   In cultures it frequently grows as a streptococcus.   When virulent, it is pathogenic to mice and rabbits, less so to guinea-pigs.   In these animals it is likely to lead to inflammations, and to rapidly fatal septicemia (twenty-four to forty-eight hours).   The blood may contain great numbers of the diplococci.   It may be introduced subcutaneously or into the peritoneum, or by intravenous injection when liquid cultures are used. Its virulence is very variable. In the sputum of a case of lobar pneumonia, early in the disease, it is likely to be virulent. The virulence is best maintained by repeated inoculations into mice or rabbits.

FIG. 62.

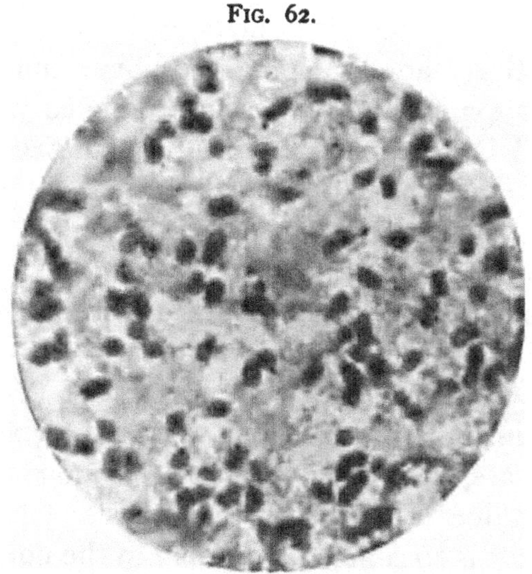

Pneumococcus showing capsule, from pleuritic fluid of infected rabbit, stained by second method of Hiss. (× 1000.)

This organism is detected very frequently in the human mouth. When taken from the mouth it is not, however, pathogenic to animals in many instances, being found virulent in only from 15 to 20 per cent. of human mouths. It is the specific cause of croupous or lobar pneumonia in man. In that disease the characteristic lesion consists of an inflammation of the lung, involving large areas, usually one or several lobes. An exudate is poured into the air-vesicles, which in the early part of the disease contains red blood-cells, imparting the rusty color to the sputum. The principal element in the exudate is fibrin. The formation of fibrin produces the liver-like consolidation or " hepatization."

The diplococci can readily be demonstrated in sections of pneumonic lung, which are best stained by carmine and gentian-violet, by the Gram method. Although the exudate at first contains many red blood-cells and the solid lung appears red, subsequently it becomes decolorized and presents a gray color. Many leucocytes will now be found to have migrated into the air-vesicles, and the lung will have become relatively anemic, instead of hyperemic. Finally, the fibrinous exudate and the cells entangled in it become softened and liquefied. Some of this liquefied exudate is absorbed into the lymphatics in the walls of the air-vesicles; part of it is expectorated.

The micrococcus lanceolatus can be detected in large numbers, sometimes almost unmixed with other bacteria, in the rusty sputum of lobar pneumonia, often showing the peculiar unstained capsule. On account of its liability to be mixed with other forms of bacteria, its presence in the sputum of cases suspected of being pneumonia is not of very great value in differential diagnosis, especially considering that it is so commonly present in the normal mouth. In a suspicious case its appearance in sputum in nearly pure culture may be significant.

Cultures from the blood of cases of pneumonia, where a large amount of blood is taken, have shown the presence of the pneumococcus in a considerable proportion of the cases, especially when severe or fatal.

The micrococcus lanceolatus is often also the cause of broncho-pneumonia and of meningitis. It produces inflammations in other situations as well, the most important being pleuritis, pericarditis, endocarditis and arthritis. The micrococcus lanceolatus may produce pseudomembranous inflammation[1] and also ordinary suppuration, although not very commonly.

[1] Cary and Lyon, *American Journal Medical Sciences*, Vol. 122, 1901.

G. and F. Klemperer claim to have obtained toxins from cultures of the pneumococcus, and to have established immunity in animals with the development in the blood of antitoxic substances. Similar attempts have been made by Washbourn and others, but the interpretation of them at the present time is not clear. An agglutination reaction has been described as occurring with the pneumococcus, but it does not yet appear to have any practical value in diagnosis.

Organisms related to the pneumococcus have been described under the names of pseudopneumococcus[1] and streptococcus mucosus.[2]

The organism named by Rosenbach, *micrococcus pyogenes tenuis*, is probably only a variety of the pneumococcus.

Micrococcus melitensis.—A micrococcus found by Bruce in cases of Malta fever. It is a round or slightly oval organism, about .5 μ in diameter, occurring singly, in pairs or in short chains. It is usually said to be non-motile, though flagella have been described. It is stained by ordinary aniline dyes, but not by Gram's method. It grows slowly, even in the incubator, and more slowly at ordinary temperatures. In gelatin the growth is feeble; there is no liquefaction. On agar pearly white growths appear after three or four days. Bouillon becomes turbid, with a sediment later. On potato there may be slight invisible growth.

Malta fever occurs chiefly about the Mediterranean. It has been observed in India, in the Philippine Islands and in Porto Rico.

It is a chronic febrile disease, accompanied by pains in the joints and perspiration, and not very fatal. At autopsies the organisms may best be recovered from the enlarged spleen. Accidental infection in man has occurred from pure

[1] Richardson, *Journal Boston Society of Medical Sciences*, Vol. V., 1901.

[2] Howard, *Journal Medical Research*, Vol VI, 1901.

cultures on a number of occasions. The disease may be reproduced in monkeys by inoculation with pure cultures. An agglutination reaction occurs in this disease. The diagnosis is best made by applying this test to the blood-serum of the patient, with a known pure culture of micrococcus melitensis.[1] A suspension of an agar culture is made in normal salt solution. The diluted serum is added so as to secure a dilution of about 1 to 50, but the dilutions used have varied widely. Precipitation quickly occurs. According to Craig the test may be made on a slide, examining with the microscope as for the typhoid bacillus (see Serum-test for typhoid fever).

Diplococcus intracellularis meningitidis.[2]—Found in the exudate of cerebro-spinal meningitis by Weichselbaum; a micrococcus about the size of the common pyogenic cocci; grows in pairs or fours, more often in pairs consisting of two hemispheres separated by an interval which does not stain; usually found within the pus-cells, in which respect it resembles the gonococcus. It is stained by ordinary methods with the aniline dyes, and is decolorized by Gram's method. It does not grow at the room temperature but only in the incubator; gelatin is not available. There is no growth on potato and scanty growth on agar or in bouillon. The development is most abundant upon Löffler's blood-serum, when round, white, shining, viscid-looking colonies with sharp outlines may be seen in twenty-four hours. The serum is not liquefied. Upon agar, or better upon glycerin-agar, the colonies are flat, round, translucent, viscid-look-

---

[1] Musser and Sailer, *Philadelphia Medical Journal,* December 31, 1898, July 8, 1899; Strong and Musgrove, *Ibid.,* November 24, 1900; Curry, *Journal Medical Research,* Vol. VI., 1901.

[2] The writer is indebted for the brief statement, which it is possible to give here, chiefly to the exhaustive Report to the Massachusetts Board of Health by Councilman, Mallory and Wright, 1898. The photograph was made from a preparation kindly furnished by Dr. Mallory.

ing, under the low power having a yellowish-brown color. The organism should be transplanted to fresh media frequently, as it rapidly loses its power of reproduction. Many of the tubes inoculated with the original material or with pure cultures show no growth.

It is moderately pathogenic for guinea-pigs and rabbits when inoculated into the pleura or peritoneum. Meningitis and encephalitis have been produced in the dog and goat by inoculation in the meninges.

FIG. 63.

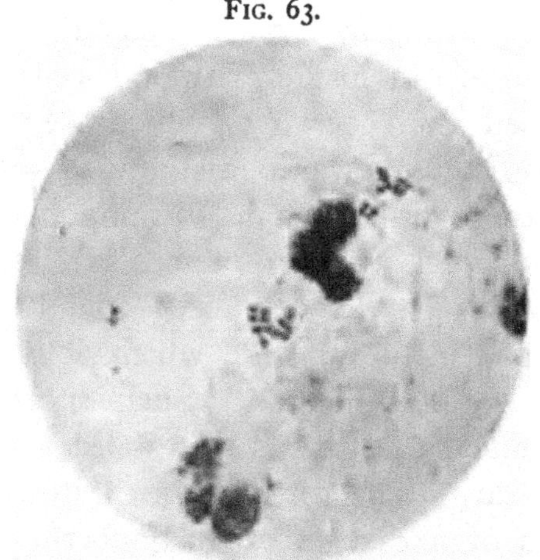

Diplococcus intracellularis meningitidis and pus-cells. (× 1000.)

This organism appears to be the principal if not the only cause of epidemic cerebro-spinal meningitis. The lesion consists of a purulent inflammation of the pia and arachnoid, extending into the brain substance, over the cord, and along the nerves. General invasion of the tissues of the body seems not to occur, but focal areas of pneumonia may be present. Spinal puncture in the lumbar region is recommended as a means of diagnosis. The puncture should be made early, and the fluid should be examined with the microscope and by cultures.

Micrococcus gonorrheæ (Gonococcus of Neisser).—
Found in pus in cases of gonorrhea. The micrococci generally are in pairs, occasionally in groups of four. The
cocci are flattened, the flattened sides facing each other, and
they are often compared to a pair of biscuits. The long
diameter of the pair of biscuit-shaped elements is about
1.25 μ. The organisms are usually found attached to the
epithelial cells or inside of the pus-cells; they are also
found in smaller numbers floating free in the fluid. They
stain with ordinary aniline dyes, for example Löffler's
methylene-blue, but not by Gram's method.

The occurrence, (1) inside of the pus-cells, (2) of pairs
of biscuit-shaped micrococci (3) which are not stained by
Gram's method, will serve to distinguish the gonococcus
from all the other ordinary pus-forming bacteria. There
are other diplococci (pseudo-gonococci), probably non-
pathogenic, which have rarely been found in the vulvo-
vaginal tract and in the urethra, which, it is said, are also
decolorized by Gram's method. Such organisms are not
likely to present all the points mentioned as characteristic
of the gonococcus. The recognition of the gonococcus in
the discharges of a case of acute gonorrhea is usually a
matter of the greatest possible ease. It must be admitted,
however, that in cases having chronic discharges, when its
detection is most to be desired, the diagnosis may become
very difficult and is frequently impossible, except by culture-
methods, owing to secondary infection with the ordinary
pus-forming or other bacteria, which may be present in
larger numbers than the gonococci themselves.

The gonococcus grows only in the incubator, and cannot
therefore be cultivated upon gelatin. Its cultivation is in
fact a matter of some difficulty. The medium usually
selected is a mixture of agar with human blood-serum. The
blood-serum from the placental blood or pleuritic or peritoneal transudates, or hydrocele fluid, has been taken. The

addition of human urine, sterilized by filtration through porcelain, to the mixture of blood-serum and agar improves its character according to some writers. A convenient medium is one consisting of one part of human serum derived from a pleuritic effusion, added to two parts of a 2 per cent. nutrient agar. The agar has previously

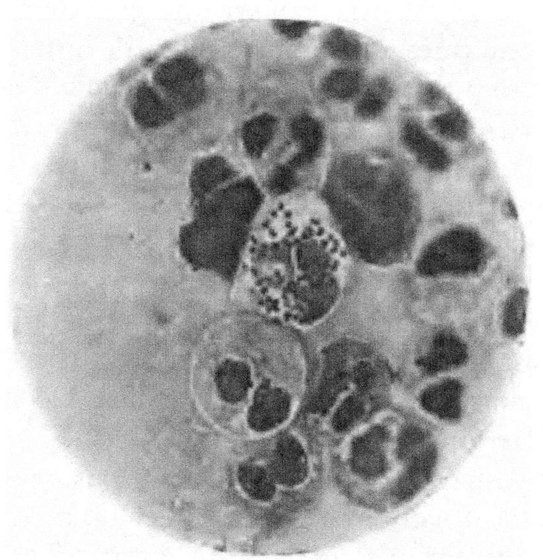

FIG. 64.

Gonococci and pus-cells. (× 1000.)

been sterilized; the two are mixed in tubes while fluid; they are cooled while in an inclined position, and are sterilized between 65° and 70° C. by the fractional method on six consecutive days. They are afterward tested in the incubator for two days.

The colonies of the gonococcus are very small, grayish-white, circular, translucent; appearing after from twenty-four to forty-eight hours. They may attain a diameter of 1 to 2 mm. The gonococcus will occasionally develop on ordinary glycerin-agar or Löffler's blood-serum medium, but the growth is likely to be feeble and cannot be relied on. The cultures live for a considerable time if kept from drying. The gonococcus is not known to produce urethritis or conjunctivitis in any of the lower animals. In the peritoneum it may cause suppurative inflammation in mice and guinea-pigs. Reproduction of the disease in man has been effected by experimental inoculation with pure cultures. Besides being the cause of gonorrheal urethritis and infection of the cervix uteri, the gonococcus has been isolated from cases of vaginitis in little girls, and from gonorrheal conjunctivitis. It has been found to be the cause of many cases of pyosalpinx, as well as of gonorrheal proctitis, arthritis, myocarditis and endocarditis; these conditions complicating gonorrhea may also be secondary or mixed infections.

Bacillus of Soft Chancre (of Ducrey).—A small, oval bacillus, usually occurring in chains. It stains with ordinary aniline dyes, but not by Gram's method. It has been cultivated on human blood-agar (also rabbit blood-agar; the medium deteriorates in a few weeks, Davis). It is cultivated with difficulty. It is found in the pus of soft chancre or chancroid, usually mixed with other organisms. It has been demonstrated in sections of the ulcers. There seems to be uncertainty with respect to its occurrence in buboes. Ducrey was able to secure it in pure culture by successive inoculations on the human skin. Although this bacillus has not yet been sufficiently studied there seems little doubt that it is the cause of soft chancre.[1]

Bacillus pneumoniæ (of Friedländer), or *Bacillus mucosus capsulatus*.[2]—A short bacillus with rounded ends, sometimes growing out to a greater length; sometimes occurring in pairs; surrounded by a capsule which is only seen in preparations made from the tissues of infected ani-

[1] Davis, *Journal Medical Research.* Vol. IX., 1903.

[2] Howard, *Philadelphia Medical Journal,* February 19, 1898; Curry, Howard, Perkins, *Journal Experimental Medicine,* Vols. IV. and V.

mals, and is not demonstrated in cultures. This bacillus is not motile. It does not form spores. It stains with the ordinary aniline dyes, but does not stain by Gram's method. It is aërobic and facultative anaërobic. It may be cultivated at ordinary temperatures, but grows better in the incubator. It does not liquefy gelatin. Stick-cultures in gelatin develop especially at the point where the puncture enters the surface of the gelatin, making what is called a " nail-shaped " growth; the growth in gelatin is white; in old cultures the gelatin acquires a brown color. It develops also on the other media. Dextrose and lactose are fermented by it; in cultures on potato, gas is formed; milk is not coagulated. It does not produce indol.

The thermal death-point is about 56° C. It is pathogenic for mice, less so for guinea-pigs and rabbits. This bacillus is sometimes found in the healthy mouth and nose. It has been known to cause inflammation, especially in the vicinity of the mouth, nose and ear, broncho-pneumonia, and more rarely empyema and meningitis. It was described by Friedländer as the specific cause of lobar pneumonia. Subsequent investigations indicate that it is comparatively seldom found in pneumonia.

There are various capsulated bacilli (capsule bacilli of R. Pfeiffer and others) which closely resemble the bacillus of Friedländer, and at least belong to the same group. The bacillus of ozæna, which has often been found in that disease is very similar. B. lactis aërogenes and B. coli communis also have many points in common with the Friedländer bacillus.

Bacillus of Rhinoscleroma.—A short bacillus with rounded ends, often united in pairs, also growing to a greater length; surrounded by a capsule; not motile; stained by the ordinary aniline dyes. It is much like the bacillus of Friedländer, but some writers have said that it is not so easily decolorized by Gram's method; this may be doubted, however. The organism has been cultivated. It is faculta- tive anaërobic. It grows rapidly, best in the incubator. It does not liquefy gelatin; its growth in gelatin stick-cultures resembles the bacillus of Friedländer. It grows on the ordinary media. Gas may be developed upon potato.

It is pathogenic for mice and guinea-pigs, less so for rabbits. Its virulence is less than that of Friedländer's bacillus.

It has been obtained from the tissues of cases of rhino- scleroma. Rhinoscleroma is a disease characterized by a chronic tubercular thickening and swelling of the skin around the nose and similar swelling of the nasal mucous membrane, sometimes followed by ulceration. It is com- monest in Austria and Italy. It has been seen in America only with the greatest rarity.

The organisms may be stained in the diseased tissues, but their detection is a matter of considerable difficulty, and they are not always found. It is not yet certain that they are the cause of rhinoscleroma.

Bacillus pyocyaneus.—A slim bacillus with rounded ends. It is motile. It does not form spores. At 56° C. it is killed in ten minutes. It is decolorized by Gram's method. It is aërobic; grows well at ordinary tempera- tures; liquefies gelatin, and grows on the ordinary culture- media. Cultures present a blue or green color, especially in transparent media. This color is not confined to the growth itself, but a blue or green fluorescence spreads over the whole medium. In old agar-cultures the color may become very dark. The pigment forms in the pres- ence of oxygen, and is due, at least in part, to the pto- maine, pyocyanin. On potato the growth is usually brown. which may be tinged with green. Milk is coagulated and peptonized and an acid reaction is developed. Indol is formed in Dunham's peptone solution.   Blood-serum is liquefied.

The bacillus pyocyaneus seems to be rather widely distributed in nature; it has been found on the skin, in normal feces, also in diarrheal discharges and in dysentery.   It is the cause of the color in blue or green pus.   It has fre-

FIG. 65.

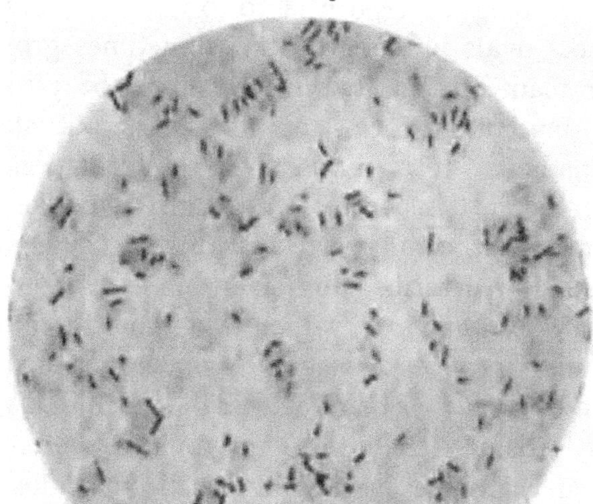

Bacillus pyocyaneus, pure culture.   (× 1000.)

quently been demonstrated in pus, but oftenest perhaps, in mixed infections.   It has been found in various abscesses, in otitis media, peritonitis, appendicitis and broncho-pneumonia.   It has been known to produce general septicemia.[1] It is pathogenic for guinea-pigs and rabbits, in whom it may produce septicemia.   In animals it may lead only to local suppuration, from which they may recover, being made im-

[1] Lartigau, *Philadelphia Medical Journal*, September 17, 1898; *Journal Experimental Medicine*, Vol. III., 1898; Perkins, *Journal Medical Research*, Vol. VI., 1901.

mune to subsequent infection with this organism. It appears that an antagonism exists between the products of the bacillus pyocyaneus and the anthrax bacillus. Rabbits which have been inoculated with cultures of the anthrax bacillus may recover if they are injected shortly after with a culture of the bacillus pyocyaneus.

Bacillus proteus.—A bacillus with rounded ends, varying much in length, breadth .4 to .6 μ; frequently appearing as short ovals like micrococci; sometimes growing out into long filaments, so that it is said to be pleomorphic. Rounded involution forms occur. It is not stained by Gram's method. It is motile. Spore formation has not been observed. It is aërobic and facultative anaërobic. It grows rapidly at ordinary temperatures. This organism was originally described by Hauser as three different species—*proteus vulgaris,* which was said to liquefy gelatin rapidly, *proteus mirabilis,* which liquefied gelatin slowly, and *proteus Zenkeri,* which did not liquefy gelatin. It seems probable that these organisms were, in fact, varieties of the same species, now called bacillus proteus. Upon gelatin-plates the colonies present a characteristic phenomenon, when seen under the low power, in the projection of processes which subsequently change their form and position, and which may become entirely detached from the original colony, so that the surface of the gelatin may become covered with so-called "swarming islands."

The proteus grows on the usual media tending to produce foul odor, decomposition and alkaline reaction. In urine it converts urea into ammonium carbonate.

This organism is one of those which were formerly described under the name of bacterium termo. It is among the most common and widely-distributed bacteria. It has been found in decomposing animal and vegetable substances, in the feces, in the urine in cystitis, and in the discharges of children having cholera infantum. It appears that this organism may occasionally be pathogenic to man, causing pus-formation, peritonitis, and even general infection.[1] Cultures injected in considerable amounts may be pathogenic to animals.

Bacillus of Bubonic Plague.—An oval or short rod-shaped bacillus, with rounded ends, sometimes possessing a capsule. It is not motile. It does not form spores. With the aniline dyes the ends stain more deeply than the middle,

FIG. 66.

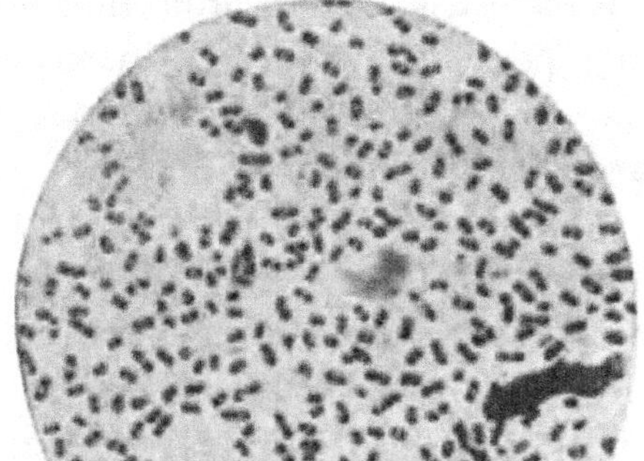

Bacillus of Bubonic Plague. (Yersin.)

called polar staining; by Gram's method it is decolorized. It is aërobic. It grows at ordinary temperatures, but better in the incubator. It grows on most media. The growths are grayish-white. Gelatin and blood-serum are not liquefied. In bouillon, the medium remains clear, while a granular deposit forms on the sides and bottom of the tube. In bouillon to which a few minute drops of sterile oil, as cocoanut oil, have been added, a growth takes place from the under side of the oil drops. Such growths extend down, and are called stalactite growths. The stalactites break off, with the slightest disturbance.

[1] Ware, *Annals of Surgery*, Vol. XXXVI., 1902.

On agar containing 3 per cent. of common salt remarkable involution forms appear. The stalactite growths and the forms occurring on salt-agar are considered the most characteristic cultural tests.[1]

It is sometimes sensitive to drying, but may survive prolonged drying. It is killed in three to four hours by direct sunlight, when spread in thin layers; in a few minutes by steam at 100° C., and in one hour by 1 per cent. carbolic acid.[2] It is pathogenic to rats, mice, guinea-pigs, rabbits, and a number of other animals.

In man it appears usually to enter through wounds of the skin. Other possible avenues of infection are the air passages, the mouth, and the gastro-intestinal tract. Plague is usually regarded as having three different possible forms, —the bubonic, the pneumonic and the septicemic. The *bubonic* form is commonest. The point in the skin at which the inoculation takes place seems generally to exhibit no inflammatory reaction. The lymph-nodes are generally swollen, especially the deep inguinal and axillary nodes. The swollen lymph-nodes may suppurate. The suppurating nodes often are infected simultaneously with micrococci. The bacilli are numerous in the enlarged lymph-nodes, but may be detected in the other organs of the body and in the blood. Fluid drawn from the buboes with a hypodermic needle may be examined microscopically, by cultures and by inoculation into rats or guinea-pigs. In the *pneumonic* or pulmonary form the bacilli occur in the sputum, and may be tested in the same manner. This type of the disease is said to be very fatal. In the *septicemic* form no primary bubo is found, or a bubonic case may become septicemic. This form is very fatal.

---

[1] Wilson, *Journal Medical Research,* Vol. VI., 1901.

[2] See Viability of Bacillus pestis, Rosenau, Marine Hospital Service, Hygienic Lab'y, Bull. No. 4, 1901.

During epidemics of plague it has been noted that rats may die in large numbers, and plague bacilli have often been recovered from the bodies of such rats. The systematic destruction by health departments of all the rats possible is important where an epidemic is present or is feared. The same applies to mice. The agency of fleas as carriers of the bacilli has been suggested, but has not yet been proved; this is equally true as to flies.

The greatest care must be used in working with the bacillus of plague. A number of fatal results have occurred through it in laboratory investigators.

Haffkine has invented a method of protective inoculation against plague by the injection of cultures of plague bacilli which have been sterilized by heat, and a little carbolic acid added. An active immunity, which is quite lasting, it is maintained, may be secured in some days. The injection is sometimes followed by considerable constitutional disturbance. This method seems likely to be of considerable value.

Yersin and others have prepared protective sera on the same general principles used in making other sera for effecting passive immunity. It is hoped they may be useful in producing quickly a temporary immunity; and the outlook for their employment in the treatment of the disease is very encouraging.

An agglutination reaction has been described; it is not likely to be of great value in diagnosis.

The period of incubation in this disease is from two to seven days. It has occasionally appeared in civilized countries during recent times though not to a very serious extent. Among the localities of importance to us it has recently visited the Philippine Islands, California and Mexico. It has ravaged the southeastern part of Asia within a few years. In the Middle Ages, and in succeeding

23

centuries, it devastated many of the countries of Europe, where it was one of the most important of the pestilences that went in those days by the name of the " Plague." It appears to have been the disease known in English history as the " Black Death."[1]

Bacillus aërogenes capsulatus.—A thick bacillus, 3 to 6 μ in length, frequently capsulated, discovered by Welch and Nuttall. The capsules may be found in preparations from

FIG. 67.

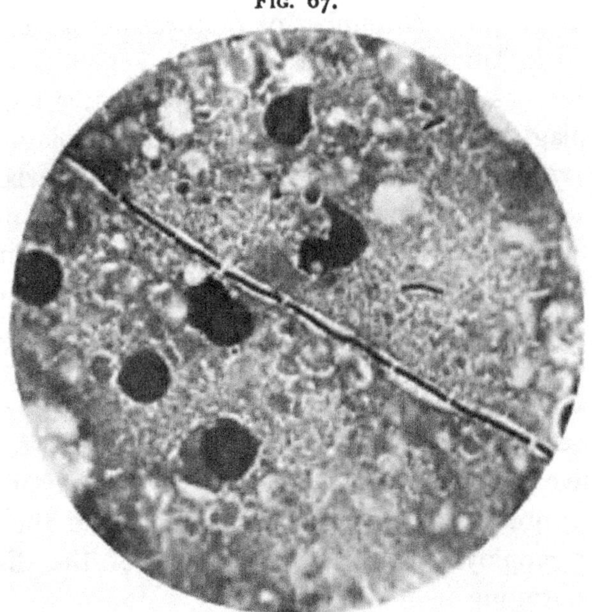

Bacillus aërogenes capsulatus, Smear-preparation from Rabbit's Liver.
(× 1000.)

animal tissues, but rarely in cultures. It sometimes forms spores chiefly in cultures on blood-serum. The vegetative forms are destroyed at 58° C. moist heat in ten minutes, but the spores withstand boiling nearly 8 minutes. It is not motile. It stains by Gram's method. It is anaërobic, and is

[1] For further details concerning plague consult articles by Barker, Novy and Flexner, *Trans. Association American Physicians,* 1902; Calvert, *American Medicine,* January 24, 1903.

readily cultivated by Buchner's method for anaërobes. It grows best at the body temperature, but will grow at the room temperature. It may liquefy gelatin slowly or not at all. The growths are whitish. In media containing lactose, dextrose, or saccharose it produces an abundance of gas; but it is also, according to Welch, able to form gas from proteids. Milk is coagulated, and the reaction becomes acid. Gas forms upon potato, where the growth is thin and grayish-white.

It occurs in the intestine of man and various other animals, in soil, sewage and water. It is not usually pathogenic to rabbits and mice. In guinea-pigs, sparrows and pigeons it may produce "gas phlegmons." It has been found on numerous occasions in the organs of human cadavers in which a development of gas had taken place, producing bubbles or cavities in the tissues, imparting to them a peculiar spongy character (German, *Schaumorgane*). Probably this is as a rule a post-mortem invasion, but there is reason to believe that in some cases it enters the circulation during life. It has been found in cases of emphysematous gangrene

FIG. 68.

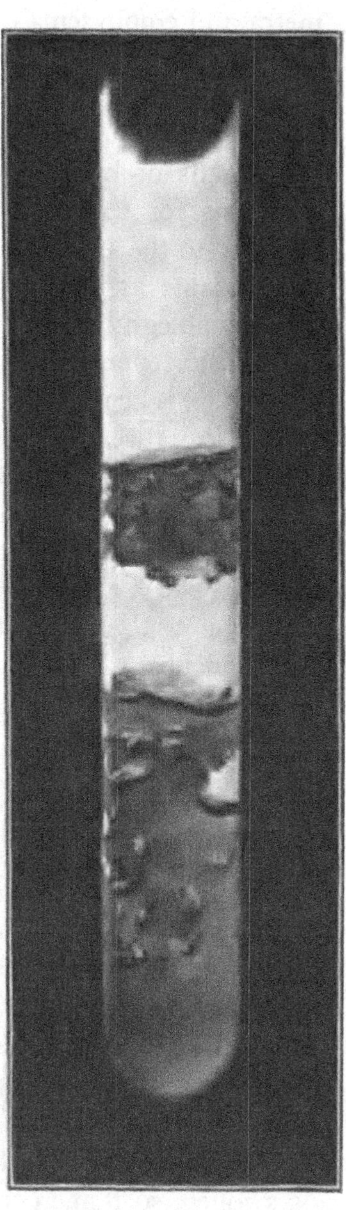

Bacillus aërogenes capsulatus, culture in dextrose-agar showing gas-bubbles.

or cellulitis, in various uterine infections, including physometra and emphysema of the uterine wall, in pneumothorax and pneumoperitonitis, and in other pathological conditions where gas occurs in the tissues. Exceptionally it may cause pus-formation.[1] This bacillus, or the gas formed by it in the organs of human cadavers, appears to have furnished the basis for some of the cases in which death has been ascribed to the entrance of air into the veins during life. It is the same as the organism described by E. Fränkel as bacillus phlegmones emphysematosæ.

Bacillus edematis maligni (French, *Vibrion septique*). —A bacillus about 1 μ in breadth, 2 to 10 μ in length, which may form threads, having rounded ends when occurring singly. It is motile, having flagella at the sides and ends. It forms spores, and may bulge at the center in consequence of the spores formed there. It is decolorized by Gram's method. It is a strict anaërobe and is best cultivated under hydrogen. It grows at ordinary temperatures, but better in the incubator. It liquefies gelatin and blood-serum. The colonies in gelatin are spherical and appear like little bubbles. It grows well upon agar. Gas may be produced in these media.

It is found in garden earth, street dirt, and in putrefying organic material. It is pathogenic to rabbits, guinea-pigs, mice, pigeons and various other animals, including man. Inoculation results in the production of swelling and edema, spreading from the point of inoculation. Gas may be produced in the tissue. It may lead to widespread septicemia.

Bacillus tetani.—A slim, straight bacillus, with rounded ends, which may form in threads. It is slightly motile. Spores form in culture-media at the end of thirty hours in the incubator. The spores are located at one end, which is swollen, so that in this stage the organism has the shape

---

[1] Welch, *Philadelphia Medical Journal*, August 4, 1900.

of a drum-stick. The spores are extremely resistant, and in the dry condition can exist for years. They are killed by moist heat at 100° C. in five minutes; by 5 per cent. carbolic acid in fifteen hours; by bichloride of mercury, 1–1000, in three hours. The tetanus bacillus stains by Gram's method. It is a strict anaërobe; it grows in an atmosphere of hydrogen, but not of carbon dioxide. It may sometimes be made to grow very well by Buchner's method.

FIG. 69.

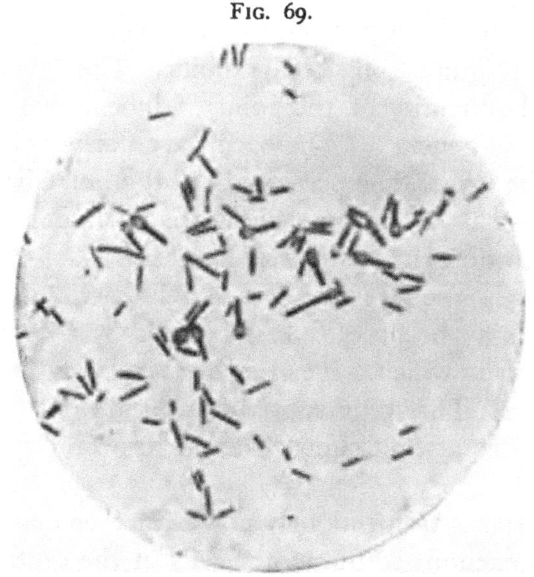

Tetanus bacilli, showing spores. (× 1000.)

It may be cultivated at the room temperature, but better in the incubator. It grows upon ordinary culture-media, preferably those containing dextrose. Gelatin is liquefied slowly; the colonies in gelatin present characteristic radiating filaments and look like a thistle. It grows on the other culture-media. Gas formation is not pronounced.

This organism appears to be widely spread in external nature, especially in the soil. It is often found in garden earth, and in the feces of herbivorous animals. McFarland believes that it may occur in vaccine virus when that is carelessly prepared, which would explain the rare occurrence of tetanus after vaccination.[1] Tetanus bacilli have been found in gelatin, and it is stated that the tetanus has followed the injection of gelatin as a hemostatic. The infection appears almost always, if not always, to be introduced through some wound.[2] Clinically, persons having the disease suffer from spasms of the muscles about the neck and the lower jaw (lock jaw). The spasms finally become general.

Inoculation with a pure culture produces tetanus in mice; also in rats, guinea-pigs and rabbits. The tetanic spasms begin in the vicinity of the point of inoculation and afterward become general. The bacilli are not widely scattered through the body; they occur only in the immediate vicinity of the original lesion, and there are no important macroscopic alterations in the internal viscera.

Tetanus is the type of the purely toxic disease. Its symptoms may be produced in animals by the injection of liquid cultures which have been deprived of their bacteria by filtration. The toxic substance appears not to be a ptomaine, as was at first supposed, and its exact nature is not determined.

The poison is tremendously powerful (see page 174). It acts as an excitant to the motor cells of the central nervous system, especially the spinal cord. Bolton and Fisch have pointed out the possibility that horses used for the preparation of diphtheria antitoxin may be infected with tetanus, and have tetanus toxin in the blood.[3]

The activity of the poison is destroyed by heat, and by direct sunlight; various chemicals diminish its intensity.

Antitoxins for tetanus have been prepared according to the principles employed for antitoxins in general. They

[1] *Journal Medical Research*, Vol. VII., 1902.
[2] Wells, " Fourth of July Tetanus," *American Medicine*, June 13, 1903.
[3] *Trans. Association American Physicians*, 1902.

have not proved very markedly successful. Unfortunately the disease is seldom suspected until a relatively large amount of toxin has formed and begun to manifest its action in the patient's body.[1]

Bacillus anthracis.—This is the largest of the pathogenic bacteria with the exception of the spirillum of relapsing fever, which is longer but more slender. The bacillus of anthrax is 1.25 μ broad, and from 3 to 10 μ long. Ba-

FIG 70.

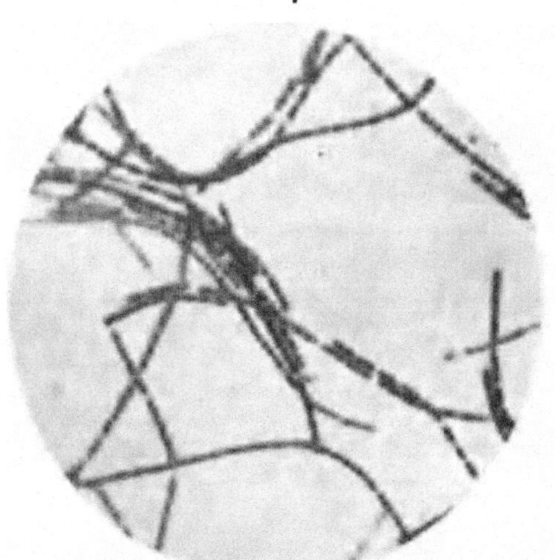

Anthrax bacilli, from a pure culture.[2]   ($\times$ 1000.)

cillus aërogenes capsulatus is of about the same size. It often forms long threads. A capsule is sometimes present. It is not motile. It forms spores, which are placed in the centers of the bacilli. The spores form only in the presence of oxygen; they do not appear in the body of an infected animal during life. Anthrax spores are the most

[1] See also Moschkowitz, *Studies Department Pathology, College Physicians and Surgeons*, New York, Vol. VII., 1899–1900. *Annals of Surgery*, 1900, p. 442.

[2] The culture was derived from a case of malignant pustule in a tanner.   The lesion was excised promptly, and the patient recovered

resistant of all pathogenic bacteria; they have been known to withstand boiling for twelve minutes,[1] 5 per cent. carbolic acid for forty days, and 1–1000 bichloride of mercury for nearly three days. The anthrax bacillus is aërobic, although not strictly so. It stains by Gram's method. It grows at the room temperature, but better in the incubator. It liquefies gelatin and blood-serum. Colonies in gelatin seen

FIG. 71.

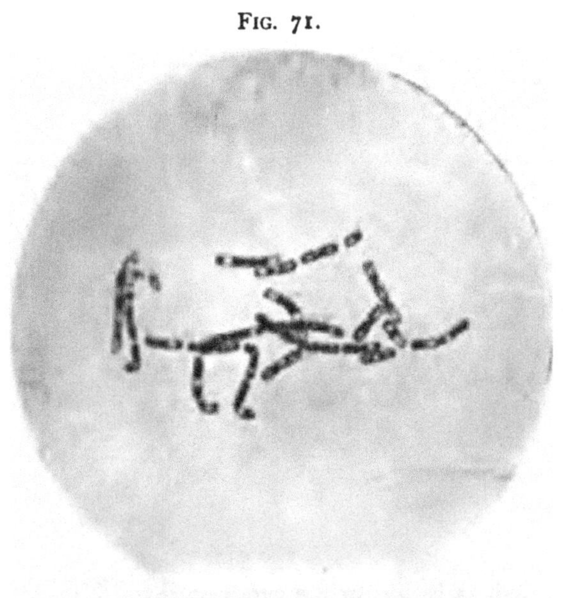

Anthrax bacilli, showing spores. (× 1000.)

under a low power display numerous, irregular, fine, hair-like projections; stab-cultures in gelatin also present fine projections passing from the needle-puncture into the solid gelatin. It grows on the ordinary culture-media; the growths are usually whitish. Cultures on potato kept in the incubator are particularly favorable to the development of spores. Milk is coagulated and later peptonized.

[1] More than half an hour. V. A. Moore, "Infectious Diseases of Animals," 1902.

It is pathogenic to mice and guinea-pigs, less so to rabbits; it is also pathogenic to sheep and cattle. Rats and pigeons are quite resistant but not entirely immune; cats, dogs and frogs are not susceptible, or but slightly so.

Anthrax is a disease which occurs chiefly in cattle and sheep. It is commoner on the continent of Europe and in Siberia than in America. In susceptible animals inoculated with virulent cultures of the anthrax bacillus septicemia is produced. Large numbers of the bacilli are found in

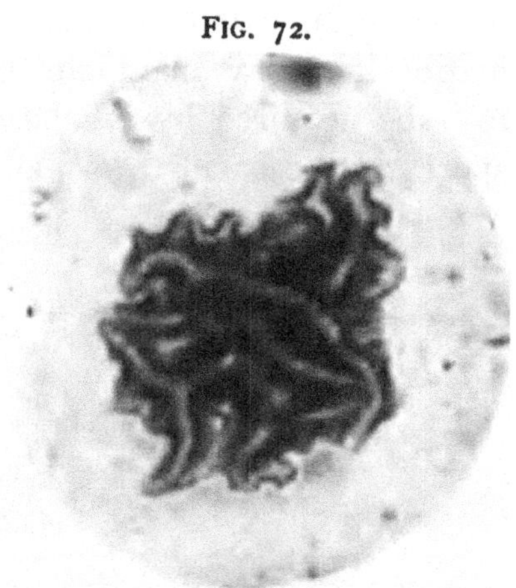

FIG. 72.

Colony of Anthrax bacilli (low power).

FIG. 73.

Bacillus of anthrax. Stick-culture in gelatin. (Günther.)

the blood, and may be crowded together in the capillaries of the liver and kidneys. Men are occasionally affected.

24

especially those whose occupations bring them in contact
with cattle or with the hides and wool of animals that die
of the disease.   The infection may enter through wounds
of the skin, where it usually produces a localized inflammation known as malignant pustule.   Anthrax of the lungs
may be acquired by inhalation of material containing the
spores of the bacilli (" Wool-sorter's disease ").   Infection
by way of the intestine oc-
curs occasionally but is less
common.   L a b o r a t o r y
workers engaged in study-
ing the anthrax bacillus
have been accidentally infected in a number of in-
stances.

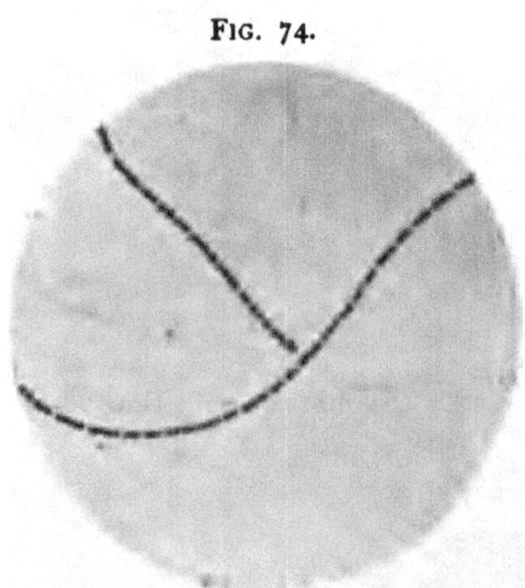

FIG. 74.

Anthrax bacilli with square or
slightly concave ends sometimes seen;
fuchsin stain.   (× 1000.)

The anthrax bacillus,
owing to its large size,
was the first of the pathogenic bacteria to be recognized, and its study has
furnished the basis for
much of our knowledge
concerning the infectious
diseases.   It was for anthrax that Pasteur developed the
idea of making a protective vaccine, shortly after he had
invented a similar vaccine for chicken-cholera.   There is
some danger attending its use.

In order to obtain material free from spores the blood of
an animal which has recently died of anthrax is taken,
because anthrax spores do not form in the living body.
Cultures made in bouillon are kept at a temperature of
from 42° to 43° C.   At this temperature spores do not
form, while the virulence of the anthrax bacillus becomes
gradually diminished.   In time the virulence is so far
diminished that rabbits will survive inoculation, and even-

tually also mice and guinea-pigs, which are extremely susceptible to anthrax. Small doses of a culture of extremely weak virulence are given to the animals which it is desired to protect, like cattle and sheep (never human beings), and subsequently a somewhat more virulent culture is employed.[1]

FIG. 75

Anthrax bacilli in the capillaries of the liver of a mouse, sketched from a section stained with fuchsin.

Bacillus influenzæ.—A small bacillus, .2 to .3 μ by .5 μ with rounded ends. It does not form spores, is not motile, and is decolorized by Gram's method. It is aërobic, grows only in the incubator, and upon media containing hemoglobin. The medium is prepared by smearing sterile blood over the surface of a tube of agar. Fresh, uncoagulated blood may, with care, be mixed with melted agar sufficiently cooled; the mixture may be poured into tubes and slanted;

[1] For details as to the results of this method, see V. A. Moore, " Infectious Diseases of Animals," 1902. For other and unique researches on immunity for anthrax see Emmerich, *Centralblatt f. Bakteriologie*, Orig. Bd. XXXII., p. 821.

the tubes should be tested in the incubator before using. The blood of some animals, as the pigeon and rabbit, may be used instead or human blood.[1] The colonies are small and transparent, looking like little drops of water, not becoming confluent.

Of a large number of bacilli, the majority are destroyed in twenty-four hours or less by drying. They die out in a similar manner in water. Experiments upon animals appear up to this time not to have been very convincing. In diagnosis, the sputum should be carefully collected in a sterile bottle. If the particles of sputum are likely to have become contaminated, rinse in sterile water. Inoculate on agar and on blood-agar. The influenza bacillus should grow only on blood-agar and have the other characters above mentioned. As far as is known, this organism grows only in man, and not outside of the human body. In cases of influenza it is found in the mucous discharges, and in the bronchi and lungs. It is the predominating organism in some cases of bronchitis.[2] According to Canon, the bacilli may sometimes be found in the blood.

Bacillus diphtheriæ (Klebs-Löffler).—A straight or slightly-curved bacillus, usually 1.2 to 2.5 μ in length, with rounded or slightly pointed ends, remarkable for showing irregularities of form, sometimes being club-shaped or spindle-shaped; branching forms have been found.[3] It is not motile, and does not form spores. It retains its color after Gram's method, but it is best stained with watery solutions of the aniline dyes, especially Löffler's alkaline methylene-blue. Very characteristic pictures are obtained by the method of Neisser:

---

[1] *Centralblatt f. Bakteriologie,* Bd. XXXII., Orig. p. 667.
[2] See Lord, *Boston Medical and Surgical Journal,* December 8, 1902.
[3] Hill, *Journal Medical Research,* Vol. VII., 1902.

SOLUTION No. 1.

Methylene-blue 1
Alcohol (96 per cent.)..................................... 20
Distilled water 950
Glacial acetic acid 50

SOLUTION No. 2.

Bismarck brown 1
Boiling distilled water.................................. 500

Stain the cover-glass preparation which has been fixed in the flame in No. 1 one to three seconds; wash in water; stain in No. 2 three to five seconds; wash in water; mount

FIG. 76.

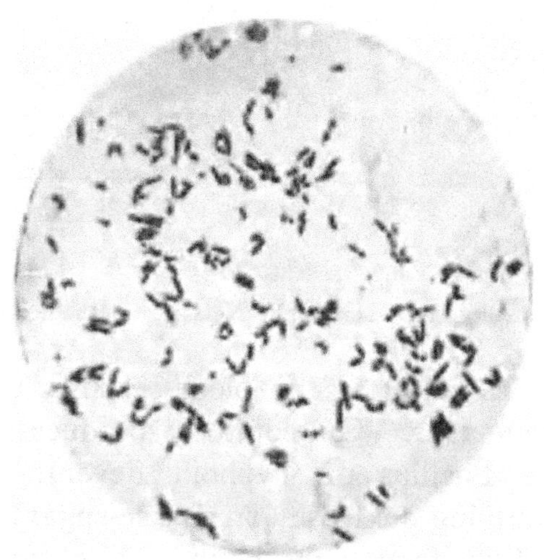

Bacillus of diphtheria.   (× 1000.)

as usual.   The body of the bacillus is stained pale brown, with dark blue spots, especially at the ends.   (Fig. 77.)

The diphtheria bacillus is peculiar in staining irregularly; certain spots stain more sharply than other portions, and darkly-stained spots are likely to occur at the ends.   It is facultative anaërobic.   It grows most rapidly in the incubator, and slowly, or not at all, below 20° C.   Gelatin is not liquefied.   It may be cultivated on various alkaline culture-media, but grows best on Löffler's blood-serum mixture. On this medium the growth consists of small white or cream-colored, slightly elevated colonies, which may become confluent. The morphology of the bacillus is most character-

FIG. 77.

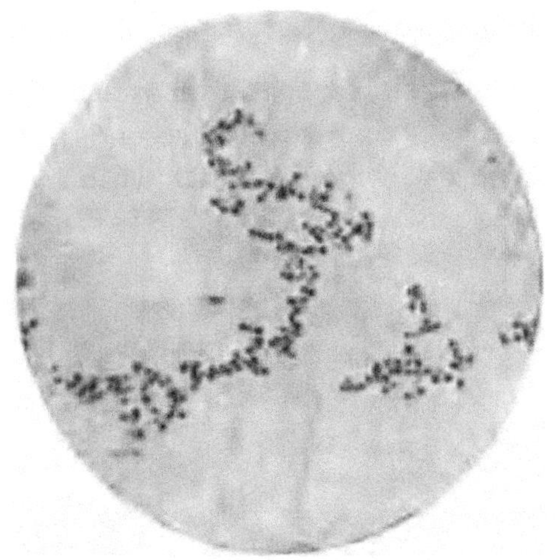

Bacillus of diphtheria stained by Neisser's method. (× 1000.)

istic when it is cultivated on blood-serum. It also grows upon glycerine-agar. On potato it produces an invisible growth (see Bacillus of Typhoid Fever). In alkaline bouillon containing dextrose (or muscle-sugar) the reaction becomes acid in forty-eight hours. The reaction of the bouillon subsequently becomes alkaline. The growth may form a pellicle over the surface of the bouillon. It has also been successfully cultivated on various media to which egg-albumen has been added.

It is killed by a temperature of 58° C. in ten minutes. It resists desiccation well.

Bacteriological diagnosis of Diphtheria.—In many large cities the bacteriological diagnosis of diphtheria is undertaken by boards of health. The methods used differ somewhat in detail, but are similar in the main, and are based upon the procedure devised by Biggs and Park for the Board of Health of New York City. Two tubes are furnished in a box. The tubes are like ordinary test-tubes, about three inches in length, rather heavy, and without a flange. Both are plugged with cotton. One contains slanted and sterilized Löffler's blood-serum mixture; the other contains a steel rod, around the lower end of which a pledget of absorbent cotton has been wound and the tube afterward sterilized. The swab is wiped over the suspected region in the throat, taking care that it touches nothing else, and is then rubbed over the surface of the blood-serum mixture. The swab is returned to its test-tube and the cotton plugs are returned to their respective tubes. The plugs, of course, are held in the fingers during the operation, and care must be taken that the portion of the plug that goes into the tube touches neither the finger nor any other object. The principles, in fact, are the same as those laid down in general for the inoculation of culture-tubes with bacteria (see page 84). In board of health work these tubes are returned to the office. When it is desirable, a second tube may be inoculated from the swab. The tubes are placed in the incubator, where they remain for from twelve to twenty-four hours, and a microscopical examination is then made of smear preparations stained with Löffler's methylene-blue. On Löffler's blood-serum kept in the incubator the bacillus of diphtheria grows more rapidly than the other organisms which are ordinarily encountered in the throat, a property which to a certain extent sifts it out, as it were, from them, and makes its

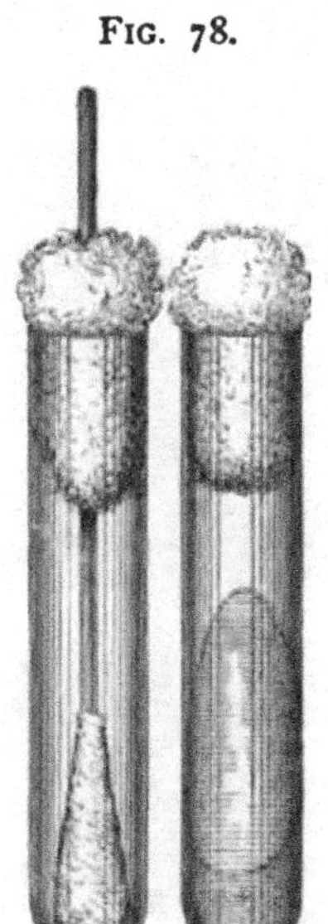

FIG. 78.

Swab and culture-tube used in the diagnosis of diphtheria.

recognition with the microscope easy in most cases. The growth, furthermore, is quite characteristic, and its nature can be predicted with considerable accuracy, even without microscopical examination, by one who has had much practice. Colonies of streptococci frequently look very like those of the bacillus of diphtheria, but these two are easily distinguished from each other with the microscope. The diagnosis of the diphtheria bacillus, then, is made from the character of the growth upon blood-serum and the microscopical examination, taking into account the size and shape of the bacilli, with the frequent occurrence of irregular forms and the peculiar irregularities in staining. In doubtful cases a second culture should be made from the throat.

The very large number of examinations that have been made by various boards of health, have shown that pseudo-membranous inflammations of the throat are sometimes caused by streptococci alone, or by other pyogenic bacteria.[1] They have also shown that the diphtheria bacillus may persist in the throat for a long

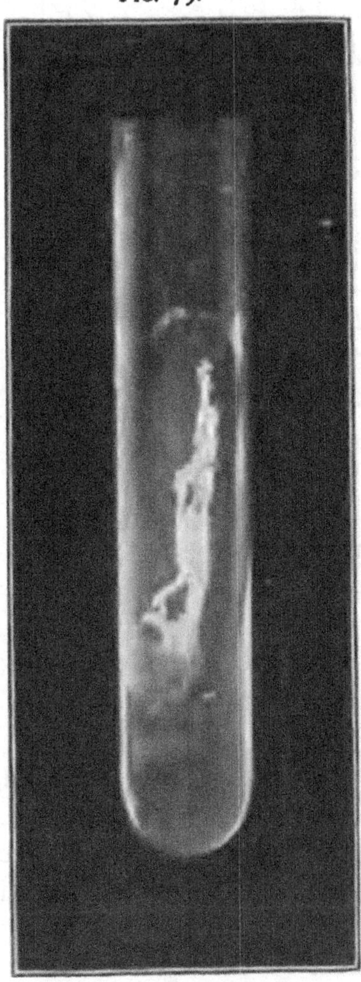

Fig. 79.

Bacillus of diphtheria, culture on glycerine-agar.

---

[1] Bissell, *Medical News*, May 31, 1902; *American Journal Medical Sciences*, February, 1903, Review of Work of Massachusetts Boards of Health.

time, occasionally several weeks after the patient has apparently recovered; also that diphtheria bacilli are occasionally found in the throat when there is an inflammatory condition without any pseudo-membrane, and that they sometimes appear in an apparently healthy throat, especially in children who have been associated with cases of diphtheria. It has been found that bacilli sometimes occur in the throat which have all the morphological and cultural properties of the diphtheria bacillus, but which are devoid of virulence when tested upon animals, *i. e.*, do not produce diphtheria toxin. Such diphtheria bacilli have frequently been called *pseudo-diphtheria bacilli*. A bacillus closely resembling the diphtheria bacillus, but without virulence, has been found in xerosis of the conjunctiva. It is called the *xerosis bacillus*. If not a transformed diphtheria bacillus, it is at least closely related. The diphtheria bacillus is subject to wide variations in morphology, so that, in dealing with unknown cultures where the forms of the bacilli are not characteristic and injection into animals is without result, it may be difficult to decide whether or not the organisms are diphtheria bacilli. Consequently another view with regard to pseudo-diphtheria bacilli has arisen. While recognizing that avirulent diphtheria bacilli occur, it is also claimed that a distinct *pseudo-diphtheria* bacillus exists, different from the diphtheria bacillus, though resembling it. It is shorter, stains more evenly, shows no polar granules by Neisser's method of staining, does not produce acid in dextrose-bouillon, and is not pathogenic to animals. It is found occasionally in the nose and throat, and has no connection with diphtheria, according to this view.[1]

[1] The different sides of this question will be found fully discussed by the following: Wesbrook, Wilson and McDaniel, *Trans. Association American Physicians*, 1900; Gorham, *Journal Medical Research*, Vol. VI., 1900, A. Williams, *Ibid.*, Vol. VIII., 1902; Denny, *Ibid.*, Vol. IX., 1903.

The diphtheria bacillus is pathogenic to animals. When it is injected into them it produces a toxemia. In the guinea-pig, which is especially susceptible, local inflammation results, and death occurs usually in two or three days. The bacilli are found to be confined to the vicinity of the wound, and not usually to be disseminated throughout the whole body. The death of the animal, therefore, is due to the poisons elaborated by the diphtheria bacilli—either poisons introduced at the original injection, or substances produced by the bacilli which may have multiplied in the animal's body. The internal viscera, especially the liver, often exhibit small areas consisting of necrotic cells; a transudation of serum takes place in the great serous cavities, and the lymph-nodes are swollen. A genuine diphtheritic membrane may be produced on the trachea of a young kitten by rubbing into it a part of a culture of the diphtheria bacillus.

As is well known, the pseudo-membranous affection produced by the diphtheria bacillus in man is generally seen in the larynx and pharynx. Membranous rhinitis is also caused by the diphtheria bacillus. On the other hand, pseudo-membranous affections of the larynx and pharynx may be produced by streptococci. Pseudo-membranes occurring in the throat during scarlet fever and measles may be due to the diphtheria bacillus, but are more often caused by streptococci. The affection known as membranous croup is usually diphtheria of the larynx, produced by the diphtheria bacillus. The diphtheria bacillus is a rare cause of puerperal fever. Although the uninjured skin is not attacked by the diphtheria bacillus, it may be present in pseudo-membranes on wounded surfaces, usually in connection with diphtheria in the throat. Most pseudo-membranes formed upon wounds of the skin are produced by other bacteria than the diphtheria bacillus, as is also the case

with the pseudo-membranous inflammations of the intestines
and bladder. Although such inflammations are often called
" diphtheritic," it must be remembered that the expression
is used in an anatomical sense, meaning that a fibrinous
pseudo-membrane has formed, extending deeply into the
tissues, which is not necessarily caused by the diphtheria
bacillus.

In cases of diphtheria in man,[1] the diphtheria bacillus is
generally found limited to the vicinity of the pseudo-membrane, and at autopsies it is not usually found in the internal viscera, excepting in the lungs, where diphtheria
bacilli may or may not be present when diphtheria is complicated with broncho-pneumonia. The general symptoms
of the disease, including the paralysis which sometimes
follows it, are due to the toxins produced by the bacilli in
the throat.

Diphtheria antitoxin. It is necessary first to obtain the toxin produced by diphtheria bacilli in a concentrated form. Virulent diphtheria
bacilli are cultivated in alkaline bouillon, in flasks plugged with cotton,
exposing a large surface to the air. The cultures are grown in the
incubator. After five to ten days they are ready, and are filtered through
porcelain. The filtrate contains the toxin. The animal usually employed is the horse, which should be healthy; the presence of tuberculosis and glanders should have been excluded, testing with tuberculin
and mallein; tetanus should also be considered, see page 272. The toxin
is injected into the horse in small doses—about 1 c.c. of the filtrate from
the bouillon culture. The dose depends on the strength of the toxin.

The injection is repeated at intervals of about one week, using larger
and larger doses, until the animal is able to tolerate a very large dose
indeed—as much as 300 c.c., or even more. If the treatment is successful the general condition of the animal should not suffer. The

[1] For a full study of the lesions of diphtheria see the Monograph of
Councilman, Mallory and Pearce, Boston, 1901.

[2] See articles by Park, A. Williams, Atkinson and T. Smith, *Journal of
Experimental Medicine,* Vol. I., p. 164; Vol. III., p. 513; Vol. IV., pp.
373 and 649; *Journal Medical Research,* Vol. IX., p. 173.

[3] W. H. Park adds 10 per cent. of a 5 per cent. solution of carbolic acid
to kill the bacilli, and filters through paper on the following day; after
adding carbolic acid the Berkenfeld filter may be used with advantage
instead of filter-paper.

injections last over a long period—usually about two or three months. The general condition of the animal remaining good, the toleration of these large doses of toxin is presumed to indicate the existence of a concentrated antitoxic substance in the blood. Small quantities of blood may be withdrawn from time to time, and the serum tested for its antitoxic strength. When a satisfactory serum has been attained, the animal may be bled and the serum saved for therapeutic purposes. Through an incision in the skin a trochar is inserted into the jugular vein. The blood is conducted into sterilized flasks with every precaution to insure sterility. The blood is allowed to coagulate and is placed for a time in the ice-chest. The serum is then withdrawn with sterilized pipettes. Small amounts of chemical germicides, as carbolic acid or chloroform, are sometimes added to assist in preserving it. This serum is the so-called antitoxin used in medical practice.

A standard to express the potency of the serum, called an immunity unit, has been devised by Behring and modified by Ehrlich. Such an immunity or antitoxic unit is the amount of antitoxic principle contained in that quantity of serum, which, when mixed and injected with one hundred times the fatal dose of toxin will preserve the life of a guinea-pig weighing 250 grams for four days.

It has been proposed to modify this test so as to lessen certain sources of error. According to the new method, one hundred fatal doses of toxin would not be used, but the amount of toxin, which, when mixed with one unit of some special and previously standardized[1] antitoxin and injected into a guinea-pig weighing 250 grams, will kill the animal in four days. This procedure would partly do away with inaccuracies, unavoidable where one hundred fatal doses of toxin are used as a standard, resulting from changes that occur in the toxin (development of toxoids, presence of toxones, etc., Ehrlich).

It has been found possible to prepare antitoxin of a high degree of concentration, so that 500 to 1,500 units may be contained in a quantity of serum which it is practicable to give at a single hypodermic injection. The large volume of statistics that have been collected from hospitals, and from physicians in private practice indicates that the use of this serum has effected a very great reduction in the mortality from diphtheria.

[1] The standard largely used in this country is an antitoxin prepared for this purpose by the *Institut für experimentelle Therapie,* Frankfort a/Main, Germany, Prof. Ehrlich, Director.

Bacillus tuberculosis.—A slim bacillus 1.5 to 4 μ in length, which very frequently presents a beaded appearance, owing to its being dotted with bright, shining spots. Branching forms have been described. The tubercle bacillus is considered by some to be a member of the actinomyces group. It is not motile. It has not been proved that spores are formed; nevertheless certain structures, like caseous lymph-nodes, have been shown to be capable of infecting guinea-pigs with tuberculosis, although tubercle bacilli could

FIG. 80.

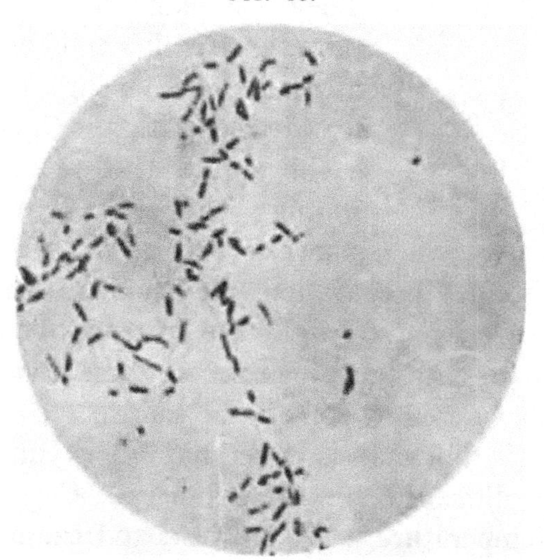

Bacillus tuberculosis, from a pure culture.   ($\times$ 1000.)

not be demonstrated in them with the microscope.   This makes it seem possible that the organisms were present as spores which eluded the microscopical examination.   The tubercle bacilli stain with the ordinary aniline dyes and by Gram's method.   As has already been stated, when stained with aniline-water dyes or carbol-fuchsin they are not readily decolorized by acids and alcohol, which fact distinguishes them from all other known bacteria excepting the leprosy bacillus, the bacillus of smegma, possibly the bacillus of syphilis (Lustgarten), and certain bacilli found in milk, butter and cow-dung and on various grasses. All of these may resist decolorization by acids or alcohol, and some resist both. They must always be kept in mind in making a diagnosis of tuberculosis. (See pages 44 and 295.) In examining sputum it is particularly important to bear in mind that acid-proof bacilli, resembling tubercle bacilli, have rarely been found in cases of gangrene of the lung. They are likely to be longer than tubercle bacilli, and branch more often, besides being less resistant to decolorization.[1] The tubercle bacilli appear to owe their peculiar staining properties to fatty substances contained in the bodies of the

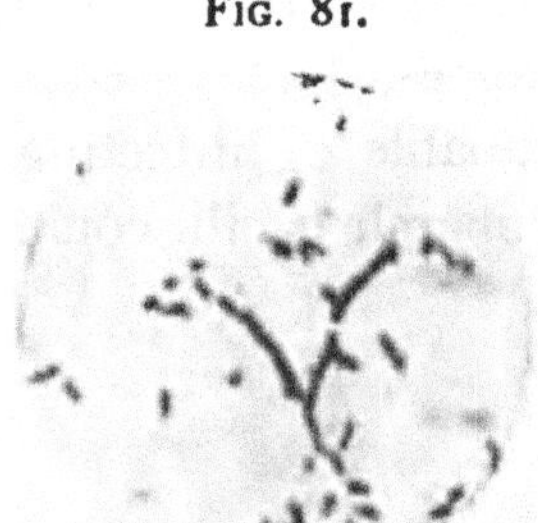

Fig. 81.

Branching form of tubercle bacillus from a culture. (× 1000.)

bacilli. In stained preparations the bacillus usually appears very distinctly beaded, owing to the presence of stained areas which alternate with unstained areas; these unstained areas have been considered by some to be spores.

The bacillus tuberculosis is aërobic. It is cultivated with considerable difficulty, best at about 38° C. It does not grow at a temperature below 29° C., and cannot therefore be cultivated upon gelatin. It grows best upon blood-serum, where the growth becomes visible in from ten to fourteen days in the incubator. It forms a dry, mealy, scaly mass, elevated above the surface, of a grayish-brown color. It also grows upon glycerin-agar; or glycerin-bouillon, on which it forms a pellicle; upon potato; upon milk containing 1 per cent. of agar and upon coagulated egg (see page 81). It is important to have the medium moist. It can be cultivated

[1] Ophüls, *Journal Medical Research,* Vol. VIII., 1902; Ohlmacher, *Journal American Medical Association,* 1901.

from tuberculous sputum only with very great difficulty. It is best to obtain it from the tissues of an animal that has died of tuberculosis, where the tubercle bacilli may be found unmixed with other bacteria. Pieces of tissue should be taken with the precautions necessary to avoid contamination, and should be broken up and rubbed over the surface of the medium. The tubes must be closed with sealing-wax, paraffin or rubber stoppers, or covered with rubber caps,

FIG. 82.

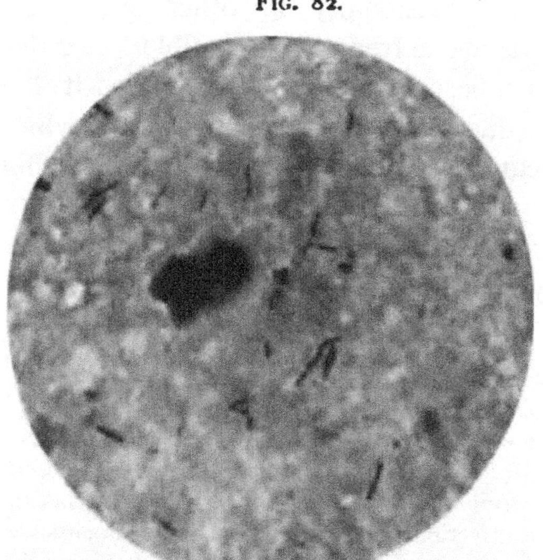

Bacillus tuberculosis in sputum, stained with fuchsin and methylene blue. Photomicrograph in two colors. ($\times$ 1000.)

to prevent drying in the incubator. If rubber caps are used they should first be left in 1–1000 bichloride of mercury for an hour, and the cotton plug should be burned before putting on the rubber cap. A number of tubes should be inoculated, using rather large particles of the tuberculous material. Among the tubes inoculated, many will fail to present any growth. After the organism has once been grown upon a culture-medium it may be propagated with less difficulty. It is best cultivated the first time upon blood-serum.

It is killed by 5 per cent. solution of carbolic acid in a few minutes. In sputum it is destroyed in twenty-four hours by a three per cent. solution of carbolic acid. It resists desiccation for months, but is killed in some hours by direct sunlight. It is destroyed in a few minutes by boiling.

It is not known to grow outside of the animal body. It is the cause of tuberculosis in man. It produces tuberculosis in apes, cows, sheep, horses, rabbits, guinea-pigs, cats, field-mice, and occasionally in other animals. Guinea-pigs and rabbits are extremely susceptible. A guinea-pig inoculated with tuberculous sputum (provided it does not die of septicemia, due to the pyogenic micrococci which are frequently present in sputum) will present a swelling of the neighboring lymph-nodes in the course of two to four weeks, and will die as a rule in from four to eight weeks, although the time may be longer.

Tuberculosis in cattle (German, *Perlsucht*) is characterized by large, nodular lesions, with a marked tendency to become fibrous, caseous and calcified. The tubercle bacilli of cattle differ somewhat from those of human tuberculosis, as was noted by T. Smith.[1] Whether or not men could be infected with bovine tubercle bacilli, has been a question that has been warmly debated in recent years. It seems to have been shown that such infection is possible; also that it is possible that cattle may be infected with human tubercle bacilli. Bovine tubercle bacilli are more virulent for some animals, as rabbits, than human tubercle bacilli.[2] It seems possible that the danger of infection from cattle has been somewhat overrated.

The lesion produced by the tubercle bacilli in the tissues of men and the lower animals is called a tubercle, which in the beginning is a grayish-white area about the size of a

[1] *Journal Experimental Medicine*, Vol. III., p. 451.

[2] T. Smith, *Medical News*, February 22, 1902; Salmon, Bureau of Animal Industry, Bul. No. 33; Adami, *Philadelphia Medical Journal*, February 22, 1902; Ravenel, *University of Pennsylvania Medical Bulletin*, May, 1902; Lartigau, *Journal Medical Research*, Vol. VI., 1901.

millet-seed. In sections of the tissue young tubercles are found to present several different structures. Near the center, one or more very large cells called giant-cells occur. They contain several or many nuclei which are frequently arranged in a crescentic manner at one side of the cell. Tubercle bacilli can sometimes be demonstrated inside of the giant-cell. Except possibly in the very youngest tubercles, a small area of necrotic tissue will always be found at the center of the tubercle.

Around the giant-cells and the necrotic area are seen large cells with distinct nuclei which resemble epithelial cells, and are often called epithelioid cells; they are also often termed granulation cells, and represent an attempt at the formation of granulation tissue. But no new-formed blood-vessels, such as are found in granulation tissue as a rule, occur in the tubercle. Tubercle bacilli may also be found among the epithelioid cells. Outside of these epithelioid cells is another layer of small cells called lymphoid cells which represent leucocytes that have appeared in this situation as a part of the inflammatory reaction excited by the presence of the tubercles. The zone of lymphoid cells may be very indistinct or wanting. Frequently it may be very difficult to make out that the cells are arranged in distinct zones at all. The cells are imbedded in a matrix consisting of the connective tissue originally belonging to the part, to which some fibrin may be added. In addition to the fact that no new blood-vessels are formed to maintain the nutrition of these newly-formed cells, the small vessels included in the tubercle and around it suffer from inflammatory changes. Owing to these causes and to a toxic substance formed by or in the tubercle bacilli, degenerative changes and necrosis take place at the central part of the tubercle. As a result of these degenerative changes the center of the tubercle becomes converted into a dry, yellowish-white, friable mass, resembling dry cream-cheese. Such material is said to be caseous, and the process is called *caseation.* Prudden and Hodenpyl found that the injection of dead tubercle bacilli into animals produced lesions having the histological characters of tubercles, but caseation did not take place.

The small tubercles first formed are called *gray* or *miliary tubercles.* As they become larger they also frequently become confluent. The larger, confluent, caseous tubercles are often called *yellow tubercles.* Swollen tuberculous lymph-nodes of the neck are among the manifestations of the condition formerly known as *scrofula.*

Masses of caseous tubercles sometimes undergo softening. In the lungs the discharge of the softened material results in the formation of a cavity. This formation of a cavity in the lungs is frequently, if not usually, accompanied by secondary infection with pyogenic micrococci. Caseous tuberculous masses may become partly calcified. Very often they may be encapsulated by new formed fibrous or scar tissue. It is possible for tuberculosis to become cured for all practical purposes by means of this process. Autopsies on human subjects have shown that such cures not rarely take place, especially in tuberculosis of the lungs occurring over a localized area. The statistics of autopsies vary widely as to the number of persons that at some time of life suffer from tuberculosis (25 or 30 per cent. and upwards). When a tuberculous area has become caseous and encapsulated and apparently quiescent, it is possible for it to be excited to renewed activity under suitable conditions, and, owing to the softening and the discharge of infected material into one of the vessels or cavities of the body, a wide-spreading and rapidly fatal tuberculosis may follow.

Tuberculosis may become disseminated throughout the body from a small focus as a starting-point. The tubercle bacilli may travel through the lymph-spaces and affect adjacent tissues, some of them reaching the nearest group of lymph-nodes. In tuberculosis of the lungs it is usual also to find tubercles in the bronchial lymph-nodes, and in tuberculosis of the intestines there is also tuberculosis of the mesenteric lymph-nodes. The disease may travel along the serous surfaces and become widely scattered throughout a cavity like that of the pleura or peritoneum. The bacilli may be expelled on some mucous surface and be carried along it to infect some point farther on, as happens when the larynx becomes infected in tuberculosis of the lungs, and when in the same disease tuberculous sputum is swallowed and leads to infection of the intestines. Finally, the infectious material may enter the blood-vessels, especially the veins, and be swept along with the blood-current to become scattered generally throughout the body. In such cases we are likely to have general or *acute miliary tuberculosis*. Almost every organ of the human body may be infected by tuberculosis. Among the most common may be mentioned the lungs, the lymph-nodes, the bones, the intestines, the skin, the meninges, and the serous membranes.

Infection, as far as we know, is always to be attributed directly or indirectly to some preëxisting case of tuberculosis in man or the lower animals. The entrance into the body is most commonly by way of the lungs, where also tuberculous disease is commonest in man, going by the name of *consumption*. This is doubtless due to the prevalent habit of expectorating in public places. Out of fifty-six samples of sputum collected in street cars by Dr. W. G. Bissell, City Bacteriologist in Buffalo, four were tuberculous. In forty-eight samples taken from the floors of a public building by Dr. C. R. Orr, of the pathological laboratory of the University of Buffalo, tubercle bacilli were found three times. According to the researches of Nuttall, a case of tuberculosis may expectorate many millions of tubercle bacilli in the course of twenty-four hours. Coughing and similar efforts may serve to disseminate the bacilli (see page 163).

Concerning the occurrence of tubercle bacilli in cow's milk and butter, see pages 148 and 149.

Cases have been recorded in which the disease was transmitted from the mother to the child in the uterus; how frequently this happens is uncertain. It is usual to attribute greater importance to an inherited tendency to tuberculosis than to the inheritance of the tubercle bacilli themselves.

Agglutination of the tubercle bacillus is said to occur with the serum of cases of tuberculosis under certain circumstances. The reaction does not seem likely to be of practical value.

Tuberculin is made by concentrating a culture of tubercle bacilli grown in glycerin-bouillon to one-tenth of its original volume, over a water-bath, and filtering through an unglazed porcelain filter. It therefore represents the products of tubercle bacilli. It was proposed by Koch as a remedy for tuberculosis, but it has not met with great success, and is little used as a therapeutic agent. It has been found, however, of great value in the diagnosis of tuberculosis, especially in cattle. When tuberculin is injected into a tuberculous animal there results considerable general disturbance, of which the most noticeable evidence is a sudden rise in temperature, while hyperemia is excited around the tuberculous area. In a healthy subject the injection produces no reaction. There is danger attending its use, so that its application in diagnosis is practically confined to cattle.[1] As a diagnostic measure in cattle it has been found accurate in the great majority of cases. Concerning tuber-

---

[1] For details as to its use in cattle see V. A. Moore, "Infectious Diseases of Animals," 1902, p. 151.

culosis in cows, see page 148. Supposing that some curative principle exists in the bodies of the tubercle bacilli themselves which could not be procured from cultures deprived of their bacilli by filtration through porcelain, Koch has recently proposed a new form of tuberculin called " tuberculin R," which consists of an extract made from dried and pulverized living tubercle bacilli. The value of this new tuberculin as a remedy is at least doubtful, and physicians are disposed to regard it with a great deal of caution.

Immunity from tuberculosis has been attained experimentally to a certain degree. In very old cultures the virulence of tubercle bacilli sometimes becomes greatly diminished. Animals which survive injec·tions of such bacilli may afterwards withstand large doses of virulent bacilli.[1]

Acid-proof bacilli resembling tubercle bacilli have been alluded to a number of times (pages 149, 154, and 288). A number of such bacilli have been cultivated, such as those of butter and grass. Injected into animals they·may produce nodules more or less like tubercles. In these nodules they sometimes assume forms resembling the fungus of actinomycosis. The tubercle bacillus rarely shows similar forms. All the bacilli of this class, including the tubercle bacillus, sometimes show branching. It is probable that the bacilli of this group are related to the fungus of actinomycosis.[2] Similar organisms have been found in fishes, in whom they produce nodules resembling tubercles; it is quite possible that the latter organisms are tubercle bacilli, which have been modified by an altered environment. Another acid-proof bacillus has been found which is pathogenic to rats, producing lesions of the skin with nodules; the disease appears in wild rats in certain localities.

Tuberculosis of Birds.—Fowls, ducks and other birds sometimes suffer from tuberculosis due to a bacillus closely resembling the tubercle bacillus of mammals. It has similar staining properties. It sometimes grows in long, branching forms. It differs somewhat from the tubercle bacillus of mammals in its cultural properties. The liver is the organ most often affected. Guinea-pigs are much less susceptible to it than to

---

[1] Trudeau, *New York Medical Journal*, July 18, 1903. Salmon, *Philadelphia Medical Journal*, June 13, 1903.

[2] Abbott and Gildersleeve, *University Pennsylvania Medical Bulletin*, June, 1902.

mammalian tuberculosis. Rabbits are somewhat susceptible, though less so than to mammalian tuberculosis.

Pseudo-tuberculosis.—Guinea-pigs and other rodents sometimes present lesions macroscopically very similar to those of tuberculosis, in which, however, the tubercle bacilli cannot be found. These affections appear not to be tuberculosis at all, and their nature is not well understood. Several organisms have been found in them, all of which are entirely unlike the tubercle bacillus.

Bacillus lepræ (of leprosy).—A slim bacillus about 4 μ in length. It is probably not motile. It is uncertain whether or not it forms spores. It stains by the Gram and the Weigert fibrin method, and it is also colored by the methods used for staining the tubercle bacillus. It takes the dye, however, more readily than the tubercle bacillus. In stained preparations it appears very similar to the tubercle bacillus, and resembles it in having alternate colored and unstained spots. Although several observers have reported success in attempts to cultivate the bacillus of leprosy, their claims have been disputed. The results of inoculation into man and the lower animals of material coming from cases of leprosy have also been uncertain. The bacillus of leprosy has been found so constantly in the tissues of those having the disease that it is generally admitted to be the specific cause. The skin and the peripheral nerves are the parts most affected, although other tissues and the internal viscera may be involved. A granulation tissue, forming nodules and thickenings, appears in the affected parts. The bacilli are found in large numbers in the nodules, partly outside of the cells, but mostly within the cells. It is still uncertain whether or not the disease can be transmitted directly from one individual to another, in extra-uterine life, or whether it can be inherited from the parents. However, no explanation can be given for the appearance of the infection in any patient, except communication with some other case. Transmission by contact seems at any rate not to take place easily.

Bacillus mallei (of glanders).—A slim bacillus with round or pointed ends, which often shows alternate light and dark spots in stained preparations. Branching forms have been described. It is not motile. It probably does not form spores. It is decolorized by Gram's method. After staining with the ordinary aniline dyes it is easily decolorized, and on that account it is difficult to demonstrate in sections of tissues. It is facultative anaërobic. It grows at the room temperature, but better in the incubator. It grows slowly on gelatin, and does not liquefy it, or only after a long time. On agar it produces a moist, white growth, on blood-serum a yellowish or brownish growth; blood-serum is not liquefied. Milk is coagulated slowly, and the reaction becomes acid. On potato the growth is characteristic in one or two days in the incubator, becoming translucent amber-yellow, later a reddish-brown, while the surface of the potato becomes discolored.

It is killed in five minutes by a 5 per cent. solution of carbolic acid, in two minutes by 1–5000 bichloride of mercury. It may survive drying for a number of weeks.

In the horse and ass it produces the disease known as glanders, which affects the mucous membrane of the nasal cavity. When the skin is involved the disease goes by the name of farcy. In the nose, nodules appear in the mucous membrane which become necrotic, forming ulcers. They may become confluent, and may extend along the adjacent surfaces as far as the lungs. There is a profuse discharge from the nose. The neighboring lymph-nodes become involved and are swollen, and nodules may be present in the internal viscera. In the skin the nodes lying underneath the skin are called farcy-buds. Histologically the nodules consist of a granulation tissue, but they tend to break down rapidly, and the process in some respects is very like ordinary suppuration.

This bacillus is pathogenic[1] to guinea-pigs, field-mice and cats; rabbits, sheep and dogs are less susceptible or only slightly so, also white and house-mice, and hogs; cattle are immune. Men are occasionally affected, especially those who have come in contact with horses. The mucous membrane of the nasal cavity may be the part involved, or the skin, or the internal viscera. In a number of instances, workers in the laboratory have been accidentally infected.

The diagnosis of the disease is best effected by the inoculation of a male guinea-pig with the material from a case suspected of being glanders, introducing it into the peritoneal cavity (Method of Straus). In about two to three days there appears a characteristic swelling of the testicle indicating the beginning of suppuration, which presently takes place; the animal usually dies after two or more weeks. At least two guinea-pigs should be inoculated; and the test may sometimes fail, when it should be repeated on other guinea-pigs.[2]

Mallein is a product obtained from an old glycerin-bouillon culture of the bacillus mallei. The cultures are placed in a steam sterilizer for several hours, and are filtered through unglazed porcelain. The filtrate contains the products of the growth of the bacillus mallei and is of much the same character as tuberculin. Injected into animals suspected of having glanders, if it produces a local and febrile reaction, the existence of glanders is indicated. It is usually successful in the diagnosis of the disease in lower animals, especially in horses, where it has been largely employed. An agglutination reaction has been described for the bacillus of glanders.

---

[1] The statements of different writers differ considerably with regard to some of these animals.

[2] Frothingham, *Journal Medical Research,* Vol. VI., 1901.

Actinomyces bovis[1] (Streptothrix actinomyces, Ray-fungus of actinomycosis).—The morphology of this organism is quite different from that of most of the bacteria. It is sometimes considered to be a bacterium of a higher type. The organism appears in the form of threads which show genuine branching. These threads make radiating, interlacing masses. Their external ends are swollen and bulbous under certain conditions. Colonies formed in this manner, seen under moderate magnification, have a radiating appearance which has given rise to the name, ray-fungus. The club-shaped external ends are readily distinguished and the growth possesses a very distinctive form. This is the shape which the organism presents as it grows in the animal body. The club-shaped ends are generally regarded as a degenerative or involution form. Transverse divisions may

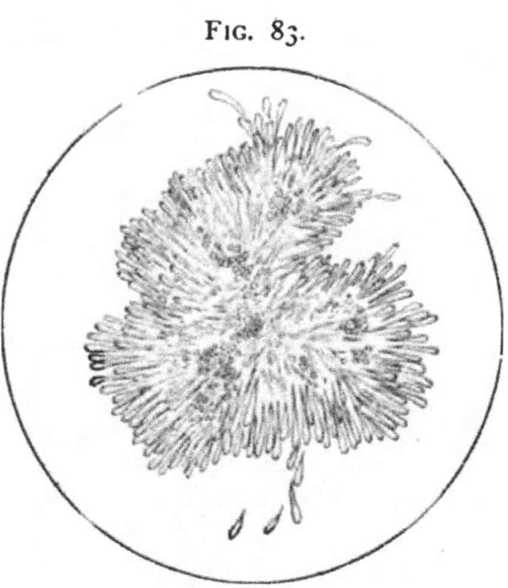

FIG. 83.

Ray-fungus of Actinomycosis. Fresh, unstained preparation from a case of lump-jaw in a cow. Diagrammatic.

sometimes be distinguished upon the threads. Spherical forms resembling micrococci may appear which may possibly be spores. In some members of this group spores form in cultures on the ends of the filaments (conidia). The organism stains with the ordinary aniline dyes, by Gram's method or the Weigert fibrin stain.

The fungus may be cultivated upon the usual culture-media, though not easily. It is facultative anaërobic. It

[1] Hektoen, *Philadelphia Monthly Medical Journal*, November, 1899; Ewing, *Bulletin Johns Hopkins Hospital*, November, 1902.

grows both at ordinary temperatures and in the incubator. The growth is not rapid. The colonies are fine, dry, elevated, irregular in form, becoming opaque. Bulbous ends upon the threads do not usually appear in cultures. The results of the injection of these cultures into the lower animals are as yet uncertain.

The disease produced by the ray-fungus is called actino-mycosis. It occurs in cattle chiefly, seldom in swine and

FIG. 84.

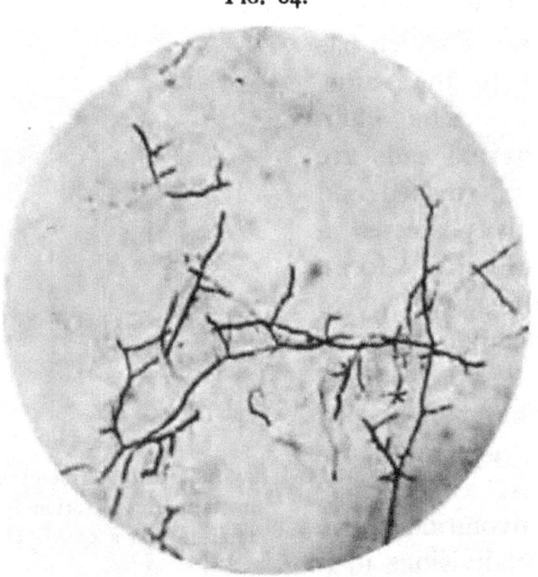

Actinomyces bovis, smear preparation from a pure culture, stained by Gram's method. (× 1000.)

horses, and occasionally in man. Infection appears to be carried by grain or particles of vegetable fiber which pene-trate the tissue. The presence of such foreign particles as well as the organism appears to favor infection. The infec-tious material frequently enters through the mouth, espe-cially in the vicinity of the teeth, but it may also occur through the skin or the mucous membranes. It leads to the formation of inflammatory, tumor-like nodules, hence the name "lump-jaw" given to the disease in cattle. Necrosis of the tissue takes place with the formation of an abscess. The pus is peculiar in containing small whitish particles which consist of little colonies of the ray-fungus, and which readily permit the disease to be diagnosed by the microscope. The material may be examined in the perfectly fresh condition without any staining. The jaw or its neighborhood is very frequently affected, or the disease may be present in other situations about the head and neck, and may involve the lungs, the intestines, and the vertebræ, ribs, and other bones. The disease is usually localized, but a number of areas may be affected simultaneously.

Besides the common actinomyces, there are numerous other ray-fungi, more or less closely related, and whose pathogenic properties are not fully determined. Generally speaking they appear to be saprophytes naturally, which occasionally become parasitic and pathogenic under especially favorable conditions. A number of species have been found in air, dust, etc., some of them chromogenic. Wolff and Israel described an anaërobic species, pathogenic to man and animals. Madura disease, Madura foot, or mycetoma is a disease occurring in India (rarely elsewhere), affecting one of the extremities, characterized by swellings, nodular deposits and abscesses. Some cases are certainly due to a member of the actinomyces group.[1]

Other branching organisms, some of them acid-proof, have been described chiefly under the name of streptothrix. In man they have been found in a variety of suppurative and necrotic lesions, in particular, broncho-pneumonias.[2]

Bacillus typhosus (Bacillus of Eberth).—A bacillus with rounded ends, varying in length, sometimes making very short, oval forms, sometimes growing out into long threads. It is very actively motile, and possesses numerous

---

[1] Compare Wright, *Journal Experimental Medicine*, Vol. III., p. 421.

[2] Norris and Larkin, *Journal Experimental Medicine*, Vol. V., p. 155; Musser, *Philadelphia Medical Journal*, September 7, 1901; Flexner, *Journal Experimental Medicine*, Vol. III.; MacCallum, *Centralblatt f. Bakteriologie*, Orig. Bd. XXXI., 1902.

flagella which arise from all parts of the surface. It does not form spores. It is not stained by Gram's method, but it may be colored with the ordinary aniline dyes, when the stain will frequently be somewhat irregular. It may be stained in sections of tissues from cases of typhoid fever, with the aniline dyes, such as Löffler's alkaline methylene-blue. It is facultative anaërobic. It grows at ordinary temperatures, better in the incubator, but grows rather more slowly than B. coli communis. Gelatin is not liquefied.

FIG. 85.

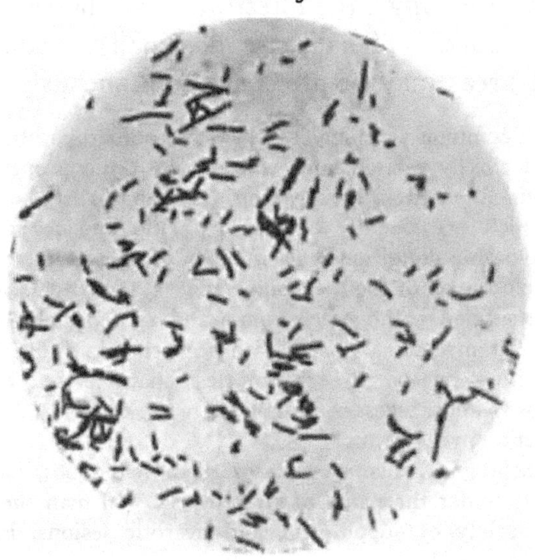

Bacillus of typhoid fever.   ($\times$ 1000.)

Young surface colonies in gelatin appear whitish, with irregular borders and more or less wrinkled surfaces, when slightly magnified. It grows on the ordinary media, and the growths are whitish. Bouillon is clouded. Milk becomes slightly acid, but is not coagulated. In media containing dextrose, acid is formed but no gas. In lactose-bouillon neither acid nor gas is formed, although when grown in milk the typhoid bacilli produce an acid reaction. The lactose-litmus-gelatin or -agar of Wurtz makes use of the blue tinge possessed by colonies of the typhoid bacillus on this medium to distinguish them from colonies of the colon bacillus and other bacteria which form acids from lactose. Neutral red has been used in the same manner, as it is said not to be altered by the typhoid bacillus, but to be changed by the colon bacillus to a yellow color. (To neutral, plain agar add 1 per cent. of a saturated aqueous solution of neutral red, and some also add .3 per cent. dextrose.)

Fig. 86.

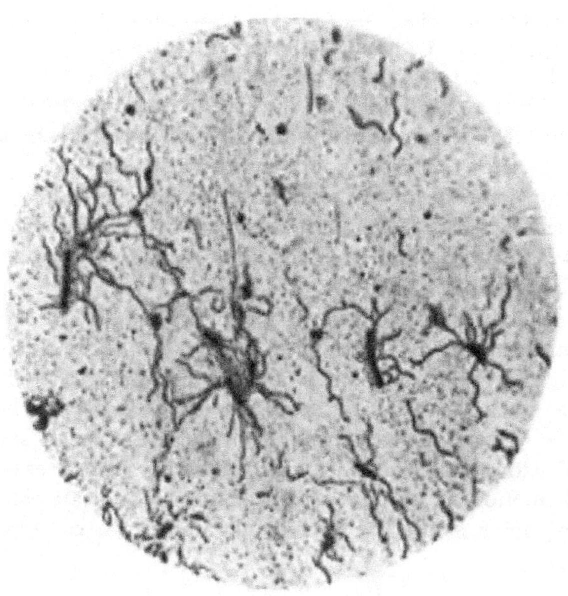

Bacillus of typhoid fever, stained by Loffler's method to show flagella.
(× 1000.)

In Dunham's peptone solution indol is not formed, as a rule. On potato it usually forms what is called an invisible growth; that is, although no development is apparent to the eye, numerous bacilli may be shown under the microscope in smear preparations made from the surface of potato inoculated about forty-eight hours previously. Occasionally a slight visible growth is seen on potato.

The typhoid bacillus is killed at 60° C. in ten minutes. It resists drying well. It can survive in soil and sewage a long time.

For a comparison of the properties of the typhoid bacillus and the colon bacillus see the latter.

Concerning the detection of the typhoid bacillus in water see page 142.

A new medium has been suggested by Hiss[1] for the isolation of the typhoid bacillus. It consists of gelatin and agar, beef-extract, sodium chloride and dextrose, and is given a slightly acid reaction. These substances are used in different proportions for plate- and for tube-cultures. This medium is of a semi-solid character, and makes use of the great motility of the typhoid bacillus in producing a uniform clouding of the medium in tubes, with the absence of a gas formation; while in plate-cultures the colonies exhibit peculiar filamentous outgrowths. It is claimed that it can be determined whether organisms are typhoid bacilli or not after thirty-six hours in the incubator.

Other special media for the identification of the typhoid bacillus have been devised by Elsner, Stoddart, by Capaldi and Proskauer, and by Piorkowski.[2] The medium of Stoddart is based upon principles similar to those applied in the medium of Hiss.

M. W. Richardson has devised an application of the serum-test to plate-colonies suspected of containing typhoid bacilli. If a typhoid colony be torn with a needle, under moderate magnification "a seething motion resembling much the appearance of a swarm of bees" may be seen. This appearance is due to the motility of the bacteria. If such a colony be touched with a small quantity of blood-serum from a case of typhoid fever, the motion is said to cease instantly and almost absolutely. Colonies of other motile bacteria do not undergo a corresponding loss of motility.

THE SERUM-TEST FOR TYPHOID FEVER.[3]

When a small quantity of a culture of typhoid bacilli is mixed with a little blood-serum derived from a case of ty-

[1] *Journal Medical Research*, Vol. VIII., 1902.

[2] Elsner, *Zeitschrift f. Hygiene*, Bd. XXI., 1895, p. 25; Stoddart, *Journal of Pathology and Bacteriology*, Vol. IV., p. 429, 1897; Capaldi and Proskauer, *Zeitschrift für Hygiene*, etc., Bd. XXIII., p. 452, 1896; Piorkowski, *Berliner klinische Wochenschrift*, 1899, p. 145.

[3] This test is often known as the "Widal reaction." For a history and general discussion of the subject see Durham, *Journal of Experimental Medicine*, Vol. V., p. 353.

phoid fever, within a few minutes the motility of the typhoid bacilli is abolished and they become agglutinated into clumps or masses. Occasionally the bacilli may eventually undergo disintegration into granular material. This reaction does not take place with the blood-serum of healthy persons or of those suffering with other diseases, nor when the blood-serum of a typhoid fever case is mixed with motile bacteria other than typhoid bacilli. It has been observed in the blood-serum of an infant born while the mother was convalescing from typhoid fever.

FIG. 87.

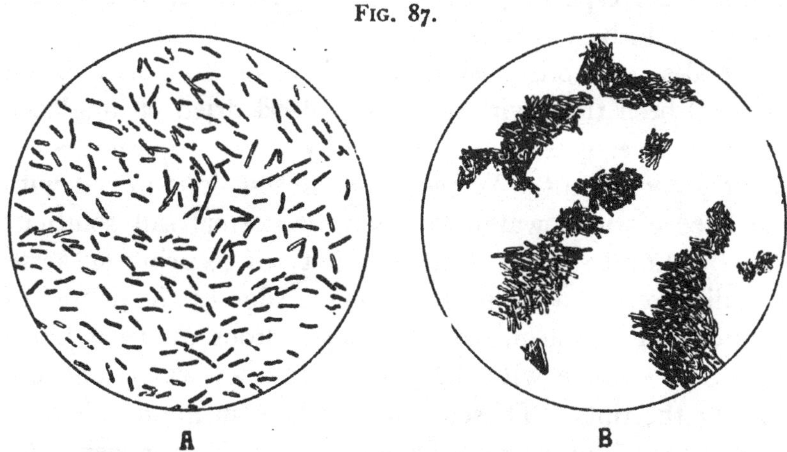

Application of the serum-reaction to typhoid bacilli. A shows the distribution of the bacilli before the reaction. It is to be remembered that they are motile and their positions may change continually. B shows clumping of the motionless bacilli after mixture with the serum of a case of typhoid fever. Diagrammatic.

The agglutinating substance has been found in blister-serum and in the milk of typhoid cases, in fluids from the serous cavities and inflammatory and edematous areas in variable amounts, and occasionally in urine, bile and tears.

The reaction may be obtained by adding blood-serum to a young bouillon-culture of typhoid bacilli kept in the incubator, when the occurrence of agglutination becomes manifest by the collection of the bacteria into visible masses or flocculi, which form a sediment. Most investigators prefer to watch the results under the microscope, using an ordinary slide, or, better, the hanging-drop. Young cultures —less than twenty-four hours old—in bouillon, and kept in the incubator, may be used, or better cultures kept at room-temperature for twenty-four hours. Johnston and Mc-Taggart recommend that the bouillon cultures be freshly made each time from stock cultures on agar, which need only occasionally be transplanted. Certain stocks of typhoid bacilli seem especially suited to this reaction, and such a stock should be secured.

Blood-serum, blister-serum, fresh blood and dried blood have all been tried with success. Blood dried on unglazed paper or cover-glasses as proposed by Wyatt Johnston is extremely convenient. To perform the test, it is mixed with sterilized distilled water, bouillon, or normal salt solution; the objection to it lies in the difficulty of securing an accurate dilution. An approximate knowledge of the degree of dilution may be acquired by mixing drops of dried blood of known volume with definite amounts of water, and observing the tints. These should be kept in mind as standards. The dilution may be measured with the hemoglobinometer or with the pipette of the hemacytometer. The New York Board of Health have found blister-serum satisfactory and easy to obtain. A little of the diluted serum is mixed on the cover-glass with a definite amount of the fresh bouillon-culture, and is examined as a hanging-drop. In a short time the characteristic clumping and loss of motility occur. At the same time a drop of the culture alone, and a drop of the culture mixed with *normal serum,* similarly diluted, should be examined as controls. The dilutions used vary from 1 part of serum in 30 to 1 in 50. The higher dilutions are more accurate. The time within which the reaction occurs varies from a few minutes to one or two hours. With little dilution the time should be short; with greater dilution it may be longer. Both clumping and paralysis of motility should take place. In a positive case the reaction should be distinct. Normal blood sometimes exhibits agglutinative properties in some degree. If the reaction in any case is not satisfactory it should be tried with a higher dilution, 1 to 50, and the result should be positive if the case is a genuine case of typhoid fever.

The reaction usually appears between the seventh day and the end of the third week of the disease; it may be seen earlier; it is often delayed and appears late. The test frequently has to be repeated when the first result is doubtful or negative. Reports indicate that the method is a great aid in the diagnosis of typhoid fever, though not infallible.

Considerable experience is necessary to acquire the judgment needed in using this test.

The agglutinating power becomes lessened after recovery, and usually is wanting at the end of a year. Rarely it may be present for a longer time, a fact that is to be borne in mind in diagnosis.

Typhoid bacilli have frequently been obtained from the stools of cases of the disease, but they are isolated only with considerable difficulty. At autopsies they are best cultivated from the spleen, in which, however, it is to be remembered, the bacillus coli communis may also be present. Puncture of the spleen with a sterilized hypodermic needle, during life, has also been resorted to as a means of diagnosis. The drop of fluid withdrawn may be examined by culture-methods for typhoid bacilli. There is probably some danger to the patient attending this procedure. Cultures made from the blood, where several c.c. are taken show that a few bacilli occur in the blood in a large proportion of cases

27

of the disease, and probably in a majority. Typhoid bacilli frequently appear in the urine (in about twenty per cent. of all cases) and the examination of urine for them has been used in diagnosis. The bacilli often occur in the gall-bladder. They have been found associated with gall-stones, and have been supposed to be one of the causes for the formation of gall-stones.[1] They may remain present in the gall-bladder or in the urine[2] long after convalescence from the disease. They have been demonstrated in the " rose spots " on the abdomen. They may be present in the lesions of the pneumonia, which frequently complicates typhoid fever, and may appear in the sputum.

Inoculation experiments in animals have not been very satisfactory. With a few exceptions, possibly, anatomical lesions resembling those of typhoid fever have not been produced by the inoculation of typhoid bacilli into animals. The injection of cultures into animals may produce death, but it can usually be shown to have resulted from the poisons contained in the cultures.

Typhoid fever is rare during the first two years of life. It frequently attacks young and robust men. The causes that bring about susceptibility to infection are not known.

The principal lesion in typhoid fever lies in the Peyer's patches of the lower part of the small intestines; the mesenteric lymph-nodes and spleen also are swollen. The typhoid bacillus may be demonstrated in sections of the walls of the diseased portions of the intestine. Cases are recorded in which no lesions were found in the intestines but where the typhoid bacilli were widely spread through the organs of the body, and which therefore represented typhoid septicemia.

Periostitis and osteomyelitis, which are not uncommon sequelæ of typhoid fever, may be caused by typhoid bacilli.

[1] Pratt, *American Journal Medical Sciences*, Vol. 122, 1901.
[2] M. W. Richardson, *Journal Experimental Medicine*, Vol. IV., 1899.

Ordinary suppuration may be produced by the typhoid bacillus, but most suppurative affections during or following typhoid fever are mixed infections, or are due to the ordinary pyogenic bacteria.

Typhoid fever is transmitted chiefly through the medium of water. It is sometimes conveyed by milk, green vegetables and oysters. Infection through the medium of dust, and by the hands and clothing probably occurs, but not commonly. Under certain circumstances the bacilli may be carried by flies.[1] In caring for cases of typhoid fever the stools, urine, sputum and linen should be disinfected. Persons handling the patient should wash and disinfect their hands.

The injection of typhoid bacilli which have been killed by heat has been advised (by Wright) as a preventive measure. The results appear to have been partially successful, but the method is still being actively studied.

Bacillus coli communis (Bacterium coli commune of Escherich, probably the same as Bacillus Neapolitanus of Emmerich, often called simply the *colon bacillus*. Passet described an organism under the name of Bacillus pyogenes fœtidus, from foul pus and mixed infections, which is probably the same as B. coli communis).—A bacillus with rounded ends, frequently of a short, oval form, when it may be difficult to distinguish from micrococci; often longer; often forming threads. It is slightly motile, having several flagella. It does not form spores. It stains with the ordinary aniline dyes and is decolorized by Gram's method. It is facultative anaërobic. It grows well at the room temperature, but more rapidly in the incubator. It does not liquefy gelatin. In gelatin plates the surface colonies are of a bluish-white color; the centers are denser

[1] Vaughan, *Philadelphia Medical Journal*, June 9, 1900.

than the borders, which are translucent. It usually grows more rapidly in gelatin than the bacillus of typhoid fever. Its growths in other media are mostly whitish. Bouillon becomes clouded. Nitrates are reduced to nitrites. In peptone solution it forms indol. On potato it forms an abundant visible growth from cream-color to pale brown.

FIG. 88.

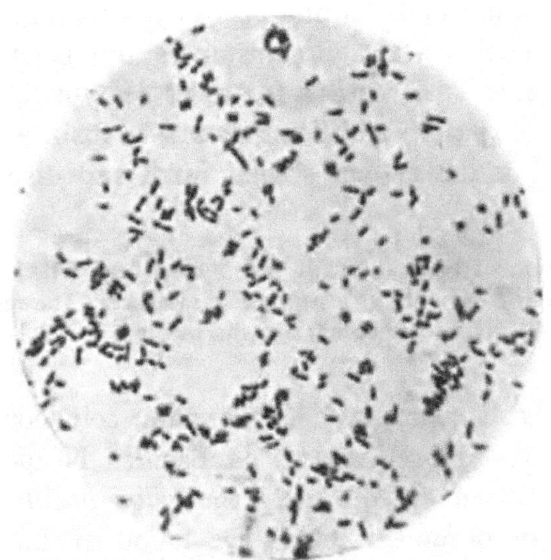

Bacillus coli communis.   (× 1000.)

Milk becomes acid and is usually, but not always, coagulated slowly. It causes the development of gas and acid in media containing dextrose or lactose. In media containing neutral red it is stated that the colon bacillus produces a yellow color with a green fluorescence. Differential points between the bacillus of typhoid fever and the bacillus coli communis are as follows:

1st. The typhoid bacillus is actively motile; the colon bacillus less actively, or slightly motile.

2d. The typhoid bacillus has numerous flagella which rise from all parts of the surface; the colon bacillus has a smaller number of flagella.

3d. In both, spore formation is absent.

4th. Both are decolorized by Gram's method.

5th. The colonies of the typhoid bacillus in gelatin develop more slowly than those of the colon bacillus.

6th. The appearance of superficial colonies in gelatin plates.

FIG. 89.

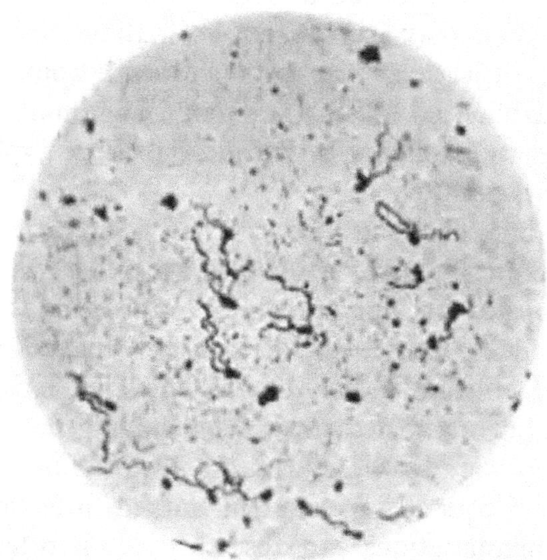

Bacillus coli communis with flagella, stained by Van Ermengem's method. (× 1000.)

7th. In media containing dextrose or lactose, the typhoid bacillus does not produce fermentation with gas and the colon bacillus does produce gas.

8th. The typhoid bacillus produces an acid reaction without coagulation in milk, and the colon bacillus produces an acid reaction and coagulation.

9th. In peptone solution the typhoid bacillus, as a rule, produces no indol, and the colon bacillus produces indol.

10th. The typhoid bacillus usually produces an invisible growth on potato, the colon bacillus a visible growth.

11th. The typhoid bacillus is said not to reduce neutral

red in media, and the colon bacillus to change it to a yellow color.

To these may be added the growth of the two organisms on special media like those of Wurtz, of Elsner and of Hiss, and the application of the serum-reaction.

Injections of cultures of the bacillus coli communis into animals produce variable and uncertain results. Subcutaneous injection may lead to pus-formation; in rabbits and guinea-pigs injections may produce death apparently from poisons introduced. With the blood of immunized animals a serum-reaction, similar to that described for typhoid fever, may be demonstrated.

Concerning the occurrence of the bacillus coli communis in the intestine of man, see page 155.[1]

At autopsies on human subjects the great viscera are often found to have been infected by the colon bacillus, usually when some lesion of the intestine existed simultaneously, but in most cases without having produced much apparent damage to the organs invaded. The bacillus coli communis frequently occurs in mixed infections, as in wounds, inflammations and abscesses. It is often found in the peritoneum in peritonitis, in the pus in appendicitis, and in the urine in cystitis; it frequently occurs in the interior of gall-stones with whose formation it may be connected.[2]

There is a large number of more or less closely-related organisms which go by the name of the *" colon group."* The limits of the colon group are extremely ill-defined.

Paracolon or paratyphoid bacilli are the names applied to certain members of the colon group which have recently been shown to be pathogenic to man. They may produce clinical symptoms resembling typhoid fever of a mild and atypical form. The affection is rarely

---

[1] See also Moore and Wright, " Bacillus coli communis in the Domesticated Animals." *American Medicine.* March 29, 1902.

[2] Lartigau, *Journal American Medical Association*, April 12, 1902.

fatal. Probably they may occur with typhoid fever in mixed and secondary infections. Characteristic lesions have not yet been observed. The bacilli have been found in the blood, spleen, liver, gall-bladder and urine. Like typhoid and colon bacilli they are motile, have flagella, are not stained by Gram's method and do not liquefy gelatin. They ferment dextrose and maltose, producing acid and gas. They do not ferment lactose. Milk at first becomes acid, later it becomes alkaline, and is not coagulated. On potato a slight visible growth occurs. Media containing neutral red become yellow, as with B. coli communis, but more slowly, and the red color sometimes returns. In respect to the fermentation of saccharose and the formation of indol reports differ; both are usually negative. The blood of the patient agglutinates the bacilli. But, as among the closely related members of this group mutual reactions are sometimes seen, this test is not to be considered invariable.[1] Several bacilli allied to the above are known. The bacillus enteridis of Gaertner is a related form which has been found in cases of meat-poisoning.

Bacillus lactis aërogenes (Bacillus aërogenes).—A bacillus having a form similar to that of the colon bacillus, described as being larger and plumper. In the main its properties are similar to those of the colon bacillus. Its colonies are more circumscribed and elevated. It is also non-motile. It coagulates milk more rapidly than the colon bacillus. It produces gas upon potato more rapidly than the colon bacillus, and more abundantly. It was described by Escherich, who also described the colon bacillus, assigning the bacillus lactis aërogenes rather to the upper part of the small intestine, and the colon bacillus to the lower portion. According to Kruse, the bacillus lactis aërogenes and its relatives differ from the bacillus coli communis chiefly in lacking motility. Like the colon bacillus it has been found many times in the urine in cystitis. See also B. acidi lactici, page 227.

Bacillus dysenteriæ (Shiga).—A bacillus with rounded ends, of the size and shape of typhoid and colon bacilli,

[1] Cushing, *Bulletin Johns Hopkins Hospital,* July–August, 1900; Strong, *Ibid.,* May, 1902; Johnstone, Hewlett, Longcope, *American Journal Medical Sciences,* August, 1902; Libman, Buxton, *Journal Medical Research,* Vol. VIII., 1902.

seldom forming threads. Most observers have found it non-motile. Vedder and Duval have demonstrated flagella. The bacillus does not form spores. It may be stained with the ordinary aniline dyes; it does not stain by Gram's method. It is facultative anaërobic. It grows at ordinary temperatures, but better in the incubator. It grows on the usual culture-media, but more slowly than B. coli communis. The growths are whitish. Colonies on gelatin plates resemble those of the typhoid bacillus. Bouillon is diffusely clouded: a precipitate may form, but no pellicle. Indol is not produced. Milk becomes acid and is not coagulated. On potato a thin pale layer forms which may become light brown. No gas is formed in media containing glucose or lactose.

Neutral-red agar is not changed. From the feces the bacillus is best cultivated on agar plates, in the incubator. Colonies of B. coli communis are often more numerous than those of the dysentery bacillus. The colonies which develop in twenty-four hours are likely to be colonies of B. coli communis. Their position may be marked on the glass with a pencil. Those which appear later are to be planted in dextrose-agar. If gas develops they are not the bacillus of dysentery; otherwise they are to be studied and identified by the cultural and other tests mentioned above, and by the agglutination reaction.

The bacilli are destroyed in a few minutes by boiling, and at 58° C. in half an hour. They appear not to be particularly resistant to the influences that are harmful to bacteria in general.

They have been found in the intestine and the discharges of acute and epidemic dysentery in various climates and countries, including the United States. Thus far their dissemination in the blood and distant organs has not been demonstrated. The lesion of this form of dysentery consists of a severe acute inflammation of the colon, frequently with necrosis of the surface and the formation of pseudomembrane. Ulceration may occur, but is usually superficial. Duval and Bassett found the bacillus of dysentery in the stools of infants having summer diarrhœa.

The introduction of pure cultures into animals by way of the alimentary canal has sometimes been followed by a certain amount of diarrhœa, but it does not appear that dysentery, as it occurs in man, has been reproduced. Most laboratory animals are, however, very sensitive to the injection into the tissues or veins of cultures, living or dead. They show the lesions produced by many toxins.

The bacillus is agglutinated by the patient's blood, but often only late in the disease and apparently not in all cases. This test seems to have only a limited value in clinical diagnosis. Many prefer to secure the reaction in a test-tube. The dilutions used vary greatly (from 1 in 20 to 1 in 100). Immunized animals develop the agglutinins in the blood. The outlook for a curative serum is encouraging.

It now seems that the bacillus of Shiga has numerous close allies, constituting with it a " group." To what extent the others of the group may be concerned in the causation of diarrhœal diseases, or may occur in the normal intestine is uncertain. According to W. H. Park some of these form indol and develop acid from mannit, which the bacillus of Shiga does not; they also differ from it in their agglutination reactions.[1]

Spirillum choleræ (Comma bacillus of cholera).—A rod- or staff-shaped organism, somewhat curved, and with

[1] Shiga, *Centralblatt f. Bakteriologie*, Bd. XXIV., 1898; Flexner, *Philadelphia Medical Journal*, September 1, 1900; Vedder and Duval, *Journal Experimental Medicine*. Vol. VI.; Gay, *University of Pennsylvania Medical Bulletin*, November, 1902; Duval and Bassett, *American Medicine*, 1902, Vol. IV., p. 417; Park and Carey, *Journal Medical Research*. Vol. IX., 1903; Strong and Musgrove, *Journal American Medical Association*, Vol. XXXV., 1900, p. 498

pointed ends, hence the name "comma" bacillus. The curved forms, placed end to end, may produce an S-shaped body. The length is from .8 to 2 μ and the breadth from

FIG. 90.

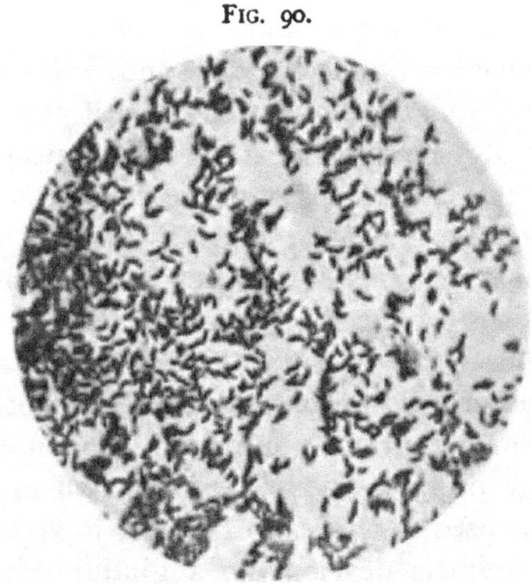

Spirillum of cholera.   (× 1000.)

.3 to .4μ. In cultures, genuine spirilla may be seen. In the whitish particles found in the stools of cases of cholera the organisms may be present in very large numbers. In these particles they may exhibit a very curious arrangement, lying parallel with one another, and, as remarked by Koch, they resemble a school of fish moving up stream. Involution forms, irregular in outline and staining poorly, are often seen in old cultures. The organism is motile, having a flagellum at one end. It does not form spores. It stains with the ordinary aniline dyes, but not by Gram's method. It is aërobic. It grows at the room temperature, but better in the incubator. On the ordinary media the growths are whitish. It grows best on neutral or alkaline media, and is very sensitive to a small amount of acid. It liquefies gelatin. The colonies on gelatin plates have a very characteristic appearance. They are nearly round at first, and granular as seen under the low power of the microscope; but at the end of about twenty-four hours the outline is slightly irregular, and the surface looks as though it were covered with finely-broken glass. The outline later becomes still more irregular or scalloped. As liquefaction of the gelatin takes place a funnel-shaped depression is formed, into which the colony sinks. The plates should be kept at a temperature of

FIG. 91.

Involution forms of the spirillum of cholera. (Van Ermengem.)

from 20° to 22° C. In stab-cultures in gelatin a white growth forms around the stab, and at the end of about thirty-six to forty-eight hours a funnel-shaped depression occurs at the surface, owing to the liquefaction of the gelatin. This depression increases in size, and the surface of the liquefied gelatin seems to be surmounted by an air-bubble, which appears to have taken the place of the part of the fluid gelatin which has evaporated. In the deeper portion of the stab liquefaction is less noticeable. The growths on agar are not characteristic. In bouillon a pellicle forms on the surface. On potato in the incubator the growth is whitish or brownish, not conspicuously elevated. After growing it in Dunham's peptone solution in the incubator, the addition of sulphuric acid develops a red color, owing to the formation of indol and nitrites, the so-called "*cholera red*" reaction.

The spirillum of cholera is said to be very sensitive to drying, and, provided the drying be complete, is usually killed within twenty-four hours. It is killed in five minutes at a temperature of 65° C. and in one hour at 55° C. It

FIG. 92.

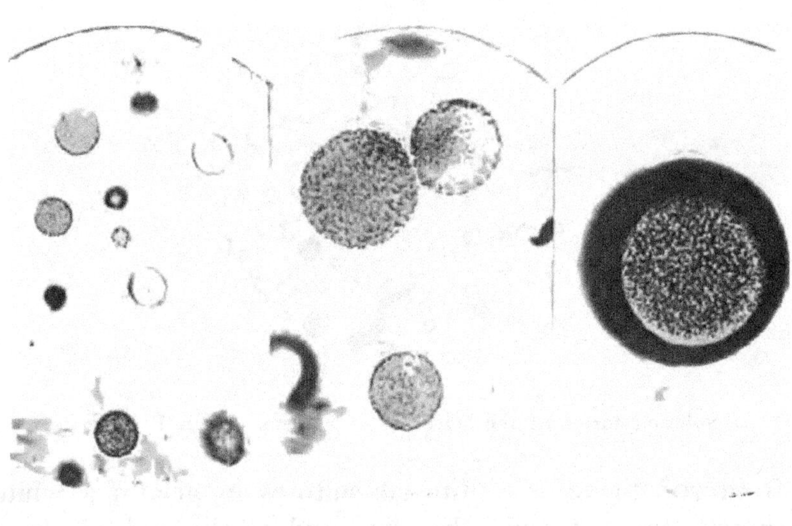

a   b   c

Spirillum of cholera, colonies on gelatin plates, × 100 to 150. (*a*) Twenty-four hours old. (*b*) Thirty hours old. (*c*) Forty-eight hours old. (Fränkel and Pfeiffer.)

may retain its vitality in water for a long time; observations vary widely in respect to determining how long. In the ordinary food-substances it may survive long enough to allow them to act as carriers of the infection if eaten raw. The important fact is that the cholera spirillum is not a strict parasite, but under favorable conditions it may maintain its vitality for some time outside of the human body.

The animals ordinarily used for laboratory experiments are, in their normal condition, not susceptible to infection with the spirillum of cholera through the alimentary canal, and no animal is known which suffers from cholera excepting man, though a disease resembling cholera can be reproduced in animals when certain conditions are complied with. In particular it is necessary to avoid the influence of the acid gastric juice.

The following plan was adopted by Koch: The gastric juice was neutralized with a solution of sodium carbonate; the movements of the intestines were quieted by the injection of 1 c.c. of tincture of opium for each 200 grams of the body-weight; and a portion of a pure culture of the cholera spirillum was introduced into the stomach. When guinea-pigs were treated in this manner, in most cases a condition closely simulating cholera was produced. The animal died with symptoms of collapse. The small intestine contained a watery, flocculent fluid in which the spirilla of cholera were numerous. The mucous membrane of the intestine was swollen and reddened.

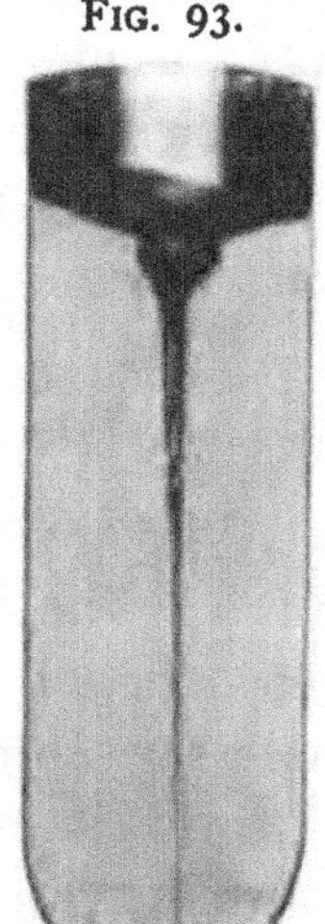

FIG. 93.

Spirillum of cholera, stab-culture in gelatin, two days old. (Fränkel and Pfeiffer.)

When mice or guinea-pigs receive an intra-peritoneal injection from a pure culture, death usually results, apparently from the toxic substances contained in the culture. Pfeiffer has shown that by repeated doses, insufficient to kill the animal, of cultures whose vitality has been destroyed by heat or otherwise, the animal may be made immune. He has also shown that when living comma bacilli are introduced in the peritoneum of an immune animal they are rapidly destroyed and disintegrated (see page 189). He has advised the use of this reaction as a means of diagnosis, inasmuch as the spirilla which apparently resemble the spirillum of cholera, but are in reality different from it, do not become disintegrated when they are introduced in the peritoneum of an animal made immune to the spirillum of cholera. It has been shown also that blood from animals made immune to cholera has an agglutinating action upon the spirillum of cholera like that seen when the blood-serum of cases of typhoid fever is mixed with living typhoid bacilli.

The outlook is encouraging for the production of a safe method for immunizing healthy persons from cholera during an epidemic.[1]

Although a positive demonstration that the spirillum of Koch is the cause of cholera is lacking, as far as the exact reproduction of the disease in animals is concerned, the necessary proof has been supplied by the accidental or intentional infection of laboratory investigators who were working with cholera, which has happened on several occasions.

Bacteriological investigations of the victims of cholera have shown that the spirilla of cholera are present in very large numbers in the watery contents of the intestine, especially early in the disease. They appear in the lumina of the glands, and they may be seen underneath the epithelial cells. They may occur in the matters vomited. They usually are not found widely spread through the organs of the body. It is probable that the symptoms of the disease result from poisonous substances produced by the spirilla or contained in them.

The infectious element in cholera is usually transmitted through water, and numerous epidemics have been studied where the infection was traced to drinking-water, and the origin of the contamination was discovered. The organ-

---

[1] Strong, *American Medicine*, Aug. 15, 1903.

isms may, however, be carried by other articles of food, and may be conveyed occasionally through contaminated clothing and bedding, and probably by flies. The excreta and bedding should be thoroughly sterilized, the hands of the attendants must be carefully disinfected. Although commoner in the summer-time, epidemics of cholera have been known to occur in the winter.

Bacteriological Diagnosis of Cholera.—When cases suspected of being cholera appear in a community, it becomes a matter of the utmost importance to determine the exact nature of the disease in order that it may not become epidemic. One of the first occasions when bacteriological methods were put into practice in the diagnosis of cholera was at the time of the appearance of that disease in the Port of New York in 1887.

According to Koch, the diagnosis may be made in twenty-four hours or less. It is important to obtain the discharges from the intestines as early in the course of the disease as possible, and while they are perfectly fresh. It may be necessary, however, to examine the moist dejecta on the linen or clothing, when no other material is available.

In the first place, one of the small, partly-solid particles which may be found in the discharges from the intestines should be smeared upon a cover-glass, fixed in the usual manner, stained with one of the aniline dyes, and examined with the microscope. If taken early in the disease, the comma bacilli may be present in large numbers, and they are likely to be arranged in more or less parallel groups (see above). If comma-shaped bacilli are thus found, a strong probability is created that the disease is Asiatic cholera. The motility of the organisms can be determined by examination in the hanging-drop. It is to be remembered that spirilla of various forms are common in the normal mouth, and may appear in the stools (see pages 152 and 228).

The diagnosis should be confirmed by the use of culture-methods. Using the small, semi-solid particles from the intestinal discharges, gelatin plates in the usual three dilutions (see page 97) should be made and kept at a temperature of 20° to 22° C. At the end of twenty-four hours or less the colonies of the spirillum of cholera should have been developed and should present the picture characteristic for these colonies in gelatin plates (Fig. 92), which enables them to be differentiated from colonies of other bacteria. From one of these colonies, preparations may be made for microscopic examination, and a set of tubes may be inoculated. The most characteristic growth will be from stick-cultures in gelatin. The growth in Dunham's peptone solution may be tested for the development of indol and nitrites.

At the time that the first smear preparations and gelatin plates are prepared, tubes of peptone solution should be inoculated directly from the intestinal contents, and kept in the incubator (Schottelius). After development has occurred, the production of indol may be tested by the addition of sulphuric acid. These tubes are especially valuable when unfavorable material or *when material containing small numbers of the spirilla is used.* In the incubator the spirilla may be expected to multiply in the peptone solution rapidly, and to appear upon the surface of the liquid in large numbers, even forming a visible film in six hours. Smears may be made from the surface part of these tubes, stained, and examined with a microscope. From the same material gelatin plates should be prepared, and examined as soon as the colonies develop.

When cultures are obtained, their effects may be tested upon guinea-pigs by injecting them into the peritoneum. The reaction described by Pfeiffer as resulting from the injection of cholera spirilla into the peritoneum of immune animals has been recommended as an additional means of diagnosis between the cholera spirillum and related forms. The agglutinating power which the blood of animals immunized to cholera has for the cholera spirillum may be employed in the same way.

In the examination of water for the spirillum of cholera, to 1 liter, or more, of water, add enough of a strong peptone solution to make it contain 1 per cent. peptone and .5 per cent. sodium chloride. (The strong peptone solution contains 20 per cent. peptone and 10 per cent. sodium chloride; is alkaline and sterile.) The water, with the peptone in it, is divided among a number of sterilized flasks. After twelve hours in the incubator, any vibrios in it are likely to have multiplied and to have formed a scum on the surface, which may be investigated for the characteristics of the spirillum of cholera according to the methods given above. See also page 142.

Since Koch's discovery of the cholera spirillum in 1883–84 a considerable number of bacteria have been described which resemble the cholera spirillum more or less closely, and which have to be taken into account in making examinations of material of any sort for it. This is particularly necessary in the investigation of water, in which such spirilla seem to occur quite frequently.

Vibrio Metchnikovi. — A comma-shaped organism, which though somewhat shorter and thicker may be very similar to the comma bacillus of cholera in form, and which, like it, may sometimes form genuine spirilla. It is motile and has a flagellum at one end. It does not form spores. It is aërobic. It stains with the aniline dyes, and is not stained by Gram's method. It grows at the room temperature. It liquefies gelatin somewhat more rapidly than the spirillum of cholera. The colonies on gelatin plates are not all alike; some of them resemble those of vibrio proteus, and others are extremely like those of the spirillum of

28

cholera. It grows upon the usual media. Blood-serum is liquefied by it. The growth on agar is grayish to yellowish, and abundant. It forms a pellicle on bouillon. In milk an acid reaction is developed with coagulation. In peptone solution it produces indol and nitrites like the spirillum of cholera. It is said to lead to the production of indol more intensely than the spirillum of cholera.

It is killed by a temperature of 50° C. in five minutes. It was discovered in chickens suffering from gastro-ente-

FIG. 94.

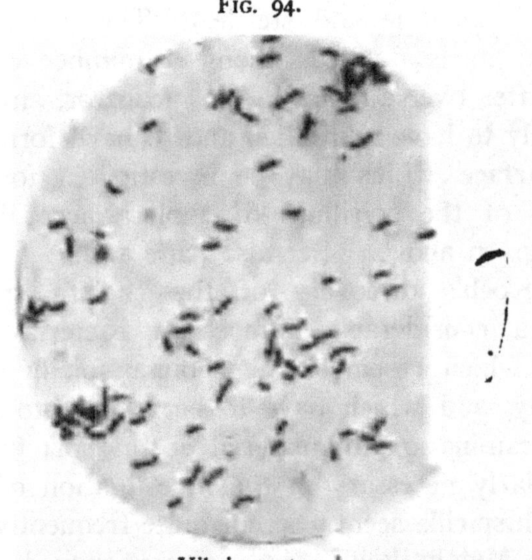

Vibrio proteus.[1]

ritis. It is pathogenic to chickens, pigeons and guinea-pigs, less so to mice and to rabbits. The comma-shaped organisms are found in the blood in guinea-pigs, pigeons and young chickens.

Vibrio proteus (of Finkler and Prior).—A comma-shaped organism somewhat larger than the spirillum of cholera, sometimes exhibiting genuine spiral forms, and also, at

[1] The magnification is a little greater than in the other photomicrographs.

times, involution forms. It is motile and has a flagellum at one end. It liquefies gelatin much more rapidly than the spirillum of cholera, and the colonies in gelatin develop more rapidly. At the end of twenty-four hours the colonies are uniformly circular, larger than those of the spirillum of cholera, and uniformly granular, when slightly magnified. On the other culture-media the growths are usually whitish. On potato it produces an abundant, moist, grayish-yellow deposit, and grows at the room temperature. It liquefies blood-serum; milk becomes acid. In peptone solution it does not form indol. It is less pathogenic to animals than the spirillum of cholera. It was supposed by its discoverers to be the cause of cholera nostras, but it appears to have no relation to that disease.

Spirillum Milleri.—A comma-shaped organism resembling vibrio proteus in many respects, and probably identical with it. In gelatin it grows more rapidly, and produces liquefaction more rapidly than the spirillum of cholera. On gelatin plates, at the end of twenty-four hours, the colonies are uniformly circular and granular, lying in little depressions resulting from the liquefaction of the gelatin. Its growths in the other media are not characteristic. It liquefies blood-serum. It does not produce indol. It is less toxic to animals than the spirillum of cholera. It was isolated by Miller from a carious tooth.

See also Spirillum sputigenum, Part III.

Spirillum tyrogenum (of Deneke).—A comma-shaped organism not so large as the spirillum of cholera. It is motile, having a flagellum at one end. It does not form spores. In cultures, genuine spirilla may develop. Gelatin is liquefied more rapidly than by the spirillum of cholera, and the colonies develop more rapidly. The circumference of the colony is round, the surface may appear somewhat granular, and it has a greenish-brown color, seen under the low power. The colonies differ noticeably from the colonies of the cholera spirillum, in the more rapid liquefaction of gelatin. Milk containing litmus becomes acid, is subsequently decolorized, and is also coagulated. It liquefies blood-serum. It does not form indol in Dunham's peptone solution. No pellicle forms in cultures upon bouillon. It is less toxic to animals than the spirillum of cholera. It was isolated originally from old cheese.

Vibrio Berolinensis.—A comma-shaped organism resembling the spirillum of cholera in form and in the position of its flagellum. It does not stain by Gram's method. It grows at the room temperature, but more rapidly in the incubator. The colonies upon gelatin, one or two days old, when magnified, are decidedly more finely granular and more transparent than those of the spirillum of cholera, and the margin is almost absolutely smooth and circular. As the colonies become older they assume a more irregular and lobulated appearance, but are still more finely granular than the colonies of the cholera spirillum. Gelatin is very slowly liquefied. Its growth on the other culture-media is not remarkable. It forms indol in peptone solution, and it increases in the upper layers of the fluid. When guinea-pigs are inoculated in the peritoneal cavity, death occurs in one to two days. This organism was discovered in the water-supply of Berlin.

Other spirilla have been isolated from water by Günther (vibrio aquatilis in Spree water) ; by Dunbar from the Elbe River; by Russell from the Gulf of Naples; by Heider from the water of the Danube Canal; and in America, by Abbott, from the water of the Schuylkill (vibrio Schuylkiliensis) ; and many others have been described to which the limits of this work will not permit of further allusion.

The **Spirillum** or **Spirochæta Obermeieri** (of relapsing fever).—A slim spirillum with numerous turns, 16 to 40 μ

in length. The ends are pointed. It is actively motile.
The spirillum is not stained by Gram's method but may be
colored by the ordinary aniline dyes. The organism has
never been cultivated. It is found abundantly in the blood

FIG. 95.

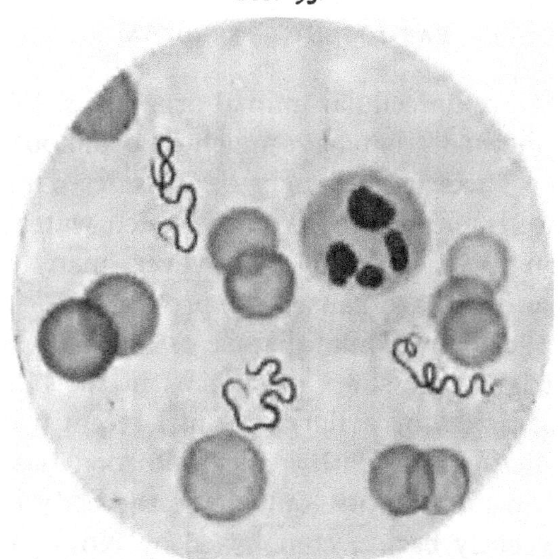

Spirillum of Relapsing Fever in the Blood. Sketched from a Stained
Specimen.

and in the spleen during the attack of fever. The spleen
is enlarged. The disease has been produced in apes by
inoculating them with blood taken from men having the
disease.

It is asserted that the spirillum is transferred by bedbugs from one person to another.[1]

[1] Karlinski, *Centralblatt f. Bakteriologie*, Bd. XXXI., Orig., 1902.

APPENDIX.

PATHOGENIC PROTOZOA.

PROTOZOA are unicellular animal organisms. As they are studied by methods that have much in common with those used for the bacteria they may be considered here briefly. Protozoa are numerous in pond and ditch water, and these species seem to be harmless. However, many diseases of the lower animals are caused by protozoa, such as surra, Texas fever, and coccidium disease of rabbits. Birds,[1] reptiles and frogs[2] may show organisms in the blood, resembling the parasites of malaria. Until recently it has been doubtful whether any pathogenic protozoön has ever been propagated in pure culture outside of the body of the host. This has recently been accomplished by Novy and McNeal for a parasite (Trypanosoma) from the blood of the rat on rabbit-blood-agar.

Amœba Dysenteriæ (Amœba Coli).—Associated with amœbic dysentery and believed to be its causative agent is the *amœba dysenteriæ,* more often named *amœba coli.* These organisms are found in the intestinal ulcers, the feces, the secondary liver abscesses and the sputum (in the latter only when an amœbic liver abscess has perforated into the lung). The lesion in the colon is a severe inflammation accompanied by necrosis chiefly of the submucous layer, and leading to extensive ulceration.[3] According to Strong, at least two distinct species of amœbæ have been found in the

---

[1] Opie, MacCallum, *Journal Experimental Medicine,* Vol. III.

[2] Langmann, *New Ycrk Medical Journal,* January 7, 1899.

[3] See Councilman and Lafleur, *Johns Hopkins Hospital Reports,* Vol. II.; see also Harris, *American Journal Medical Sciences,* Vol. 115, 1898.

feces in man, only one of which is pathogenic and the cause of dysentery. Unfortunately the designation, *amœba coli*, has been applied to both species. The amœba of dysentery should be designated *amœba dysenteriæ*, limiting the term *amœba coli* to the non-pathogenic form or forms.

The *amœba dysenteriæ* is a unicellular organism, 20–50 μ in diameter when at rest, consisting of a clear, homogeneous ectosarc and a granular endosarc, with an eccentrically placed nucleus. The endosarc contains a number of vacuoles of variable size and very frequently red blood-corpuscles, as well as other foreign bodies such as bacteria, pigment granules, etc. Many red blood-corpuscles may be seen crowded together in a single amœba. The organism is actively amœboid, extending its substance into processes or pseudopodia of varying forms. This amœboid motion assists in making easy the recognition of the parasites under the microscope and in distinguishing them from large, swollen cells found in the feces. The stool should be examined while fresh and still warm.

The non-pathogenic amœba (amœba coli), also occasionally found in the intestinal tract of man, differs from the pathogenic dysenteric organism chiefly in its much smaller size (10–24 μ) and the invariable absence of red corpuscles from its interior. The protoplasmic granules are also, as a rule, smaller and are difficult to recognize. The amœba dysenteriæ produces experimentally definite ulceration of the gut of cats, whereas the amœba coli is harmless. Both varieties of amœbæ may be stained by a special stain devised by Mallory.[2]

---

[1] Strong, Circulars on Tropical Diseases, No. 1, Chief Surgeon's Office, Headquarters, Division of the Philippines, Manila, P. I., February, 1901. *Ibid.*, No. 11, April, 1901. (Both reports may be obtained from the United States Government, Washington.)

[2] Mallory, *Journal of Experimental Medicine*, September, 1897, Vol. II., p. 529.

The Malarial Parasite.[1] (Plasmodium or Hematozoön Malariæ).—The organisms of malaria consist of at least three different species, each associated with one of the three types of malarial fever: The *tertian* parasite with benign tertian malarial fever, the parasite reaching maturity in forty-eight hours; the *quartan* parasite with benign quartan malarial fever, the cycle of development requiring seventy-two hours; and the *æstivo-autumnal* parasite with malignant, æstivo-autumnal fever, developing to maturity in a variable period of from twenty-four to forty-eight hours. The parasites are studied to best advantage in a drop of fresh, fluid blood placed between a cover-glass and slide and examined with an oil-immersion objective. For method of making and staining dry preparations, see pages 55 and 108.

Tertian Parasite.—This appears in its youngest form as a small, round, colorless, hyaline body within the red corpuscle, seen during and just after the chill of the disease. This body may be actively amœboid, suddenly changing its contour into various forms. Its size gradually increases, and fine, dark, actively-motile, dancing pigment granules begin to appear at its periphery.

The red corpuscle harboring the parasite, with the growth of the latter, becomes gradually paler and expands in size. The parasite as it grows loses its earlier amœboid movement and the pigment granules, still actively motile, accumulate. Near the end of forty-eight hours, the organism finally fills the red corpuscle, only a faint rim indicating the latter. The ripe parasite now divides it into from fifteen to twenty-five small, round, hyaline spores which are arranged somewhat radially about the pigment granules which have lost their motility and become concentrated in a clump at the center of

[1] Thayer and Hewetson, "The Malarial Fevers of Baltimore," *Johns Hopkins Hospital Reports,* 1895, Vol. V., and reprinted; Thayer, "Lectures on the Malarial Fevers," New York, 1897.

the spore-forming organism. The spores finally break apart and scatter, each destined to invade a red corpuscle and start anew the cycle of development. This cycle may be repeated over and over again, producing a corresponding number of malarial paroxysms.

FIGS. 96–99.

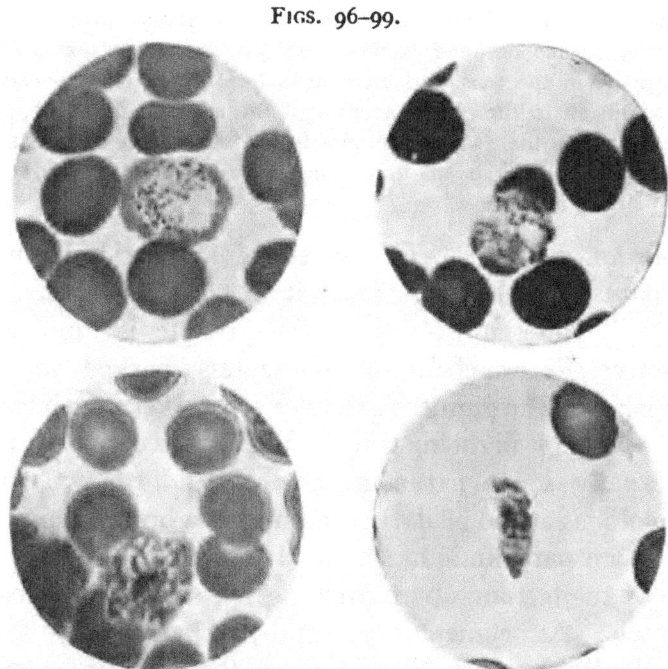

Malarial Parasites in Various Stages. (× 1000.) 96, 97 and 98 are tertian parasites; 98 shows the completion of segmentation. 99 is the crescentic form of the æstivo-autumnal parasite.

Certain full-grown parasites do not complete the cycle of development by sporulation, as described, but, breaking loose from the corpuscle, remain as "*extra-cellular*" bodies. These are seen chiefly after the paroxysm as large, round, pale bodies containing numerous dancing pigment granules scattered through their substance. They ultimately degenerate and disappear. Some of these extra-cellular forms may be seen to develop long slender processes (flagella), having a very active whip-like motion. Flagella are never observed in perfectly fresh blood but develop only after the blood has been drawn some time, usually fifteen or twenty minutes.

The extra-cellular forms of the parasite (*gametes*), incapable of further development in their human *intermediate host*, can continue their life cycle only when, by chance, they happen to be sucked into the body of a mosquito of the genus Anopheles (*definite host*), in which they undergo a second complete (sexual) cycle of development with the ultimate production of spores (sporozoids). When in turn the spores chance to be inoculated into the blood of man by the bite of an infected Anopheles, the man becomes infected, and the cycle of development in the red corpuscle, already outlined, commences. The second or sexual cycle of the parasite in the mosquito, here described for the tertian organism, applies as well to the other varieties of the malarial organism, namely the quartan and the æstivo-autumnal forms, in the case of each starting from the extra-cellular mature forms of the organism found in the blood of the human host.[1]

Quartan Parasite.—This resembles quite closely the tertian parasite, but differs from it in certain respects. The young, hyaline, intra-corpuscular parasite is more highly refractive, its amœboid motion is less marked and more sluggish, and the pigment granules are darker, much coarser, and have very slight motility. The infected red corpuscles are usually somewhat contracted instead of swollen and their color is apt to be darker, assuming a bronzed hue. The full-grown parasite is much smaller than the corresponding form of the tertian, approximating the size of a normal red corpuscle. As segmentation begins, a characteristic appearance develops which distinguishes the quartan organism, namely, the coarse pigment granules are drawn toward the center of the parasite in certain converging straight paths, giving a stellate arrangement to the pigment, until finally it becomes clumped entirely at the center in a solid mass. The segmenting forms of the quartan parasite thus present a more symmetrical arrangement of the spores, which often resemble the petals of a " marguerite." These spores are oval and number only from six to twelve, being fewer than those of the tertian segmenting parasite. The quartan

---

[1] Lyon, " The Inoculation of Malaria by the Mosquito: A Review of the Literature," *Medical Record,* February 17, 1900.

extra-cellular forms are smaller than those of the tertian, being about the size of a red corpuscle, and contain coarse pigment granules in active motility until degeneration occurs. Flagella may develop from certain extra-cellular forms. The entire development of the quartan parasite occupies about seventy-two hours.

Æstivo-autumnal Parasite.—This parasite develops to maturity in from twenty-four to forty-eight hours and is usually regarded as representing a single species, though certain observers claim to distinguish two distinct varieties. The usual description of a single variety is here adopted. The youngest forms (hyaline bodies) resemble those of the tertian and quartan organisms, but are distinctly smaller and more highly refractive. They often present a ring-like appearance. They are amœboid. Pigment granules later appear at their periphery, but are exceedingly minute and scanty, seldom more than one or two being seen. These granules have little or no motility and in fact are with difficulty made out. The hyaline bodies remain small, seldom exceeding one third the diameter of a red corpuscle. The infected corpuscle is apt to be crenated, shrunken and dark. These are the forms seen in the circulating blood in early infections; the mature forms, with the exception of the extra-cellular forms, developing in the spleen and bone-marrow, rarely reach the general circulation. Blood from the spleen shows the full-grown forms in abundance. The segmenting forms resemble those of the tertian parasite both in the numbers of the segments and in their arrangement, but are much smaller in the aggregate, as well as in the individual segments.

After the fever has lasted about one week, extra-cellular forms make their appearance in the circulating blood. These are crescentic, ovoid or small round bodies, containing coarse pigment granules at their center, generally arranged in a ring. The crescents and ovoid bodies are highly refractive and are in length about equal to the diameter of a red corpuscle, sometimes larger. The round forms are smaller than a red corpuscle, with the pigment arranged centrally in a ring. They may become flagellated after the blood has remained outside the body for some minutes. Any of the extra-cellular bodies may show remnants of the red corpuscle attached to its side, like a bib. The extra-cellular forms are concerned in the cycle of development of the organism in the mosquito, and are sterile in the human body. They are exceedingly resistant to quinine and may continue in the blood for long periods of time.

Melaniferous leucocytes (phagocytes) are seen in the blood, being especially abundant after the paroxysm in all forms of malarial infection.[1]

Small-pox and Vaccinia.—Micrococci of various sorts have been found in the pustules of small-pox and vaccinia, but indicate only a secondary infection. Other microörganisms have been described. The most important are certain bodies often considered protozoa. In both small-pox and vaccinia small round homogeneous bodies, 2 to $4\,\mu$ in diameter, have been found in the epithelial cells of the vesicles. Inoculation of vaccine lymph into the rabbit's cornea leads to the production of similar bodies in the epithelial cells of the cornea. W. Reed[2] found small amœboid bodies in the blood in cases of small-pox and vaccinia. Vaccine virus that has been filtered through the Chamberland or Berkenfeld filter is no longer active. From this it may be presumed that the organism causing it is not too small to be seen with the microscope.

Councilman, Magrath and Brickerhoff,[3] as a result of

---

[1] See also Ewing, *Journal Experimental Medicine,* Vols. V. and VI.

[2] *Journal Experimental Medicine,* Vol. II., 515; see also, Anna Williams and Flournoy, and W. H. Park, *N. Y. Univ. Bull. Medical Sciences,* II., October, 1902.

[3] *Journal Medical Research,* Vol. IX., May, 1903.

recent studies, believe that the bodies above mentioned are protozoa. Segmentation of the bodies is described, resulting in the formation of spore-like bodies. The spore-like bodies undergo a further or second cycle of development within the nucleus. The second cycle also ends in segmentation. The two cycles were seen in small-pox; in vaccinia, only the first or extranuclear bodies were observed.

YELLOW FEVER.

It has already been indicated (page 160) that the study of cases of yellow fever has failed to prove that this disease is caused by bacteria. On the other hand, evidence that it is transmitted by the mosquito, Stegomyia, has been increasing.

As malaria and some other diseases transmitted by mosquitoes are due to protozoa, careful search has been made for protozoa in yellow fever. Examinations of the blood of individuals having yellow fever have been without result. Recently Parker, Beyer and Pothier have studied specimens of Stegomyia fed on such blood. They claim to have found that these mosquitoes became infected with a protozoön parasite, a portion of whose cycle was worked out.[1] Ordinary mosquitoes did not contain the parasite. Blood-serum from a case of yellow fever was filtered through a Berkenfeld filter and injected into two healthy subjects; one subject developed yellow fever and the other did not. The interpretation of these results in connection with the experiments of Reed and Carroll (page 161) is not at present clear.

Trypanosomes.—A number of species of Trypanosoma have been described, which produce diseases in the lower animals; recently one has been stated to be the cause of disease in man.[2] The trypanosoma is a protozoön belonging to the flagellata. It is of an elongated spindle-shaped form, with a nucleus, and has a flagellum at one end, which extends along a thin edge, called the undulating membrane. It is actively motile. It occurs in the blood, between, but not in, the blood-corpuscles. Its length is two to several times the diameter of a red corpuscle. Members of this genus are the cause of surra (a fatal dis-

[1] Marine Hospital Service. Yellow Fever Institute, Bulletin No. 13, March, 1903.

[2] For a full description of the life history and classification of Trypanosoma see Salmon and Stiles, Emergency Report on Surra. U. S. Bureau Animal Industry, Bulletin No. 42, 1902. See also Francis, Marine Hospital Service, Hygienic Laboratory, Bulletin No. 11, 1903.

ease of horses and mules occurring in India and the Philippine islands) and of the tsetse-fly disease of South Africa; while others are found in rats, birds, amphibia and fishes. In the horse the infection is transmitted by the bites of flies. Novy and McNeil have succeeded in cultivating the trypanosoma of rats on rabbit blood-agar.[1]

Several cases were reported during 1902 where trypanosomes were found in the blood of individuals from tropical Africa, showing that this group of parasites may occur in man.[2] The symptomatology of these infections requires further study. Still more recently it has been claimed by Castellani that a trypanosoma is the cause of " sleeping sickness," a disease of the natives of Africa. He states that the parasites may be demonstrated in the cerebro-spinal fluid obtained by lumbar puncture and, with greater difficulty, in the blood, during life. Many cases also show at autopsy streptococcus infection, which is believed to be a secondary invasion.[3]

[1] Novy and McNeil, in Contributions to Medical Research dedicated to Victor C. Vaughan, 1903.
[2] *British Medical Journal*, May 30, 1903.
[3] *British Medical Journal and Lancet*, June 20, 1903.

INDEX.

30

Made in the USA
Monee, IL
07 July 2026

56552376R00215